图书在版编目（CIP）数据

新古典家居艺术 /（俄罗斯）普蒂洛夫斯卡亚编；

李婵，张晨译 . -- 沈阳 : 辽宁科学技术出版社，2016.5

ISBN 978-7-5381-9747-1

Ⅰ . ①新… Ⅱ . ①普… ②李… ③张… Ⅲ . ①住宅 - 室内装饰设计 - 图集 Ⅳ . ① TU241-64

中国版本图书馆 CIP 数据核字 (2016) 第 059700 号

出版发行：辽宁科学技术出版社
　　　　　（地址：沈阳市和平区十一纬路 29 号 邮编：110003）
印 刷 者：利丰雅高印刷（深圳）有限公司
经 销 者：各地新华书店
幅面尺寸：230mm×290mm
印　　张：18.5
字　　数：40 千字
印　　数：1～1100
出版时间：2016 年 5 月第 1 版
印刷时间：2016 年 5 月第 1 次印刷
责任编辑：马竹音
封面设计：周　洁
版式设计：周　洁
责任校对：周　文
书　　号：ISBN 978-7-5381-9747-1
定　　价：338.00 元

联系电话：024-23284360
邮购热线：024-23284502
http://www.lnkj.com.cn

NEO-CLASSICAL
ART IN HOME DESIGN
新古典家居艺术

（俄）玛丽娜·普蒂洛夫斯卡亚 / 编

李婵　张晨 / 译

辽宁科学技术出版社

PREFACE 前言

Everytime looking at amazing interiors of Louvre or Hermitage I ask myself – how would our life look like without such masterpiece and historical heritage? Each detail, shape, curve make us fall in love with this beauty, which was created by a human and his talent.

Is it possible to repeat it nowadays? Of course. Thanks to modern technology we can copy each detail easily. Do we want to have and enjoy such beauty in our houses? No doubt, as a wish to own the best is in our blood. Are we ready to follow our dream? This is the main question of the modern neo-classical interior.

With the help of various technologies and tools today it is possible to make the interior full of effects of flying birds, rustle of leaves, breath of wind, blossom of flowers... Getting inspiration from the historical interiors, will we create our own masterpieces or will we stay on the way of easy coping of our past and pasting it into our future – this is a relevant question in the modern world of the interior design.

If the modern interior is a space in which people don't want to stop, as the speed of their life is incredible, the modern home interior is made for people who want to relax on the sofa, have a proper dinner in the dining room and enjoy the calming atmosphere of their house. When the owner sits on the sofa, he should be surrounded by beautiful details, which he wants to examine. These details should not only set the mood – they should transform in the interior under different lighting and angles.

Today the classical interior appears in a modern way with the panoramic windows, high ceilings, new technologies, different level of comfortability and flowing space. Classical interior is built for examination and contemplation, for the owner's comfort and satisfaction. You cannot get tired from the classical interior, as each detail is made with love and warm, and that is why it is still the most popular trend in the interior world.

In particular classical interior with its many-sided choice, admiration of details and freedom, make any of our fantastic and bold ideas come true.

Will we stop being afraid to create something new and will we learn to dream? This is the question I often ask myself.

This book is a helpful guide on the best neo-classical interior ideas from different authors. Amazing interiors, collected from different parts of the world, show the huge range of our possibilities. The use of modern technologies made the interior more bright, memorable and unique. And looking at this book, published in the country which

has changed our perception of the technological progress speed, I know the answer. I am sure – we are ready for the future, and this future will be great!

Marina Putilovskaya

每当看到卢浮宫或者埃尔米塔日博物馆令人叹为观止的室内空间时，我不禁要问自己：如果没有这些伟大的杰作和历史遗产，我们的生活会是怎样？每个细节、造型和曲线都让我们爱上这种美，这种美是人类才华的结晶。

现在我们有可能复制这些杰作吗？答案当然是肯定的。有了先进的现代科技，我们可以轻易复制出每个细节。我们想在自己家里欣赏这样的杰作吗？毫无疑问！因为"想要拥有最好"的欲望永远在我们的血液中流淌。我们是否已经准备好去实现这个梦想呢？这是现代新古典室内设计面临的主要问题。

现在我们拥有各式各样的技术和工具，可以在室内空间中营造出各种效果，不论是飞鸟翱翔、落叶沙沙、微风轻拂还是繁花盛放，等等。在从历史古迹的室内空间中汲取灵感之后，我们是创作属于自己的杰作，还是仅仅简单地复制过去然后将其粘贴到未来呢？这是现代室内设计界面临的一个很有意义的问题。

如果说现代室内空间是人们不想停留的地方（因为他们的生活节奏已经快得难以置信），那么，现代家居空间却是完全不同的——人们想在沙发上休息，想在餐室享受一顿美餐，想在家里体验轻松愉悦的氛围。当主人坐在家中沙发上的时候，他的周围应该满是精致的细节，让他忍不住去仔细观赏、品味。而且这些细节不应该预设了空间的氛围；在不同的照明效果烘托下，从不同的角度来欣赏，室内空间应该呈现出不同的面貌。

如今，古典的室内设计以现代的方式表现出来——全景的开窗、高高的天花、最新的技术、不同的舒适度以及流畅的空间体验。古典风格的室内空间是让人去慢慢观赏、享受的，主人在这个过程中会体验到舒适感和满足感。在古典风格的环境中你是不会感到厌倦的，因为每个细节都是有温度的，承载了设计者满腔的心血，这也正是古典主义至今仍是室内设计界最流行的趋势的原因。

特别是在古典风格的室内设计中，设计师可以有多样的选择，对细节和自由可以无限地追求，这让我们任何异想天开的大胆想法都有可能变成现实。

我们是否会止步于复制，而不敢去创造新的东西呢？我们会有自己的梦想吗？这是我经常问自己的问题。

本书汇集了来自不同设计师的最优秀的新古典室内设计作品。这些作品分布在世界各地，充分显示了我们拥有的无限的可能性。现代科技的运用让室内空间更加耀眼、更加独特、更加令人过目难忘。这本书在中国出版，这个国家改变了我们对科技发展速度的认识。看到这样一本书，我对答案更加确定。我相信，我们已经为未来做好准备，未来将无限美好！

玛丽娜·普蒂洛夫斯卡亚

CONTENTS 目录

010 Chapter 1 Overview of Neo-classical Home Design
第一章 新古典家居设计概述

012 1. Historic Development of Home Design
1. 家居设计的发展

023 2. Key Points in Neo-Classcial Home Design
2. 新古典家居的设计要点

026 3. Design Principles in Neo-Classical Homes
3. 新古典家居的装饰原则

032 4. Modern Technologies in Neo-Classical Homes
4. 新古典家居的现代形式

034 Chapter 2. Functions of Neo-Classical Home Design
第二章 新古典家居的功能艺术

036 1. Reasonable Division of Space
1. 理性空间划分

038 *Kutuzovskaya Rivera*
库图卓弗斯卡娃里韦拉公寓

046 2. Safety and Privacy in Home Spaces
2. 家居中的安全感和私密性

048 *320m² Apartment*
320平方米公寓

056 3. Private Territories for Family Members
3. 家庭成员之间的私人领域

058 *Penthouse*
顶层公寓

064 4. Interior Passageways
4. 室内过道的处理

066 *Nikolsky Deadlock*
尼克尔斯基公寓

072 5. Ample Space and Right Zoning
5. 充足的空间和正确的形式

074 *House on the Jasmine Street*
茉莉大街别墅

080 6. Open and Enclosed Areas
6. 开敞空间与封闭空间

082	*Pokrovsk*
	波克罗夫斯克公寓

090	7. Satisfaction of Requirements
	7. 满足人的欲望

092	*Brown Residence*
	布朗别墅

102 Chapter 3 Decorations of Neo-Classical Home Design
第三章 新古典家居的装饰艺术

104	1. Space with Different Elements
	1. 不同元素组成的空间造型

106	*170m² Apartment*
	170平方米公寓

114	2. Dynamic and Luxurious Details
	2. 细节处的生动和华丽

116	*3rd Avenue*
	第三大道住宅

124	3. Mix and Match of Materials
	3. 多种材料之间的混搭艺术

126	*Valley Falls Estate*
	峡谷瀑布别墅

136	4. Diversified Colour Schemes
	4. 多重色彩的装点

138	*Romanovo*
	罗曼诺夫别墅

148	5. Furniture
	5. 家具造型与环境

150	*Zvenigorod*
	兹维尼哥罗德别墅

158	6. Application of Baroque Style in Home Design
	6. 巴洛克造型在住宅中的应用

160	*Casa Mataró*
	马塔罗公寓

166	7. French Residence in Black and White
	7. 黑白之间的法式住宅

CONTENTS 目录

168 *Twomey Country House*
图米乡间别墅

178 Chapter 4 Neo-Classical Art for a Better Living Environment
第四章 新古典家居的环境艺术

180 1. Natural Light in Every Room
1. 自然光线进入每一个房间

182 *40 Beverly Park*
贝弗利公园40号住宅

190 2. Combined Lighting
2. 家居中的复合照明

192 *Godolevski*
高德莱温斯基公寓

200 3. Ventilation
3. 家居环境的通风

202 *Round Hill Road Residence*
朗德山路住宅

210 4. Indoor Greening
4. 家居绿化设计

212 *Summerland*
夏日乐园

220 5. Connection between Interior and Exterior
5. 室内与室外的交流

222 *Sokol*
索科尔公寓

230 6. Environmental Psychology and Home Design
6. 环境心理学与家居设计

232 *Hashemian Family Residence*
哈什米安别墅

240 Chapter 5 Neo-Classical Art in Modern Homes
第五章 新古典家居的新形式

242 1. Renovation of Old Houses
1. 旧环境的再创造

244	*Ivory Requiem* 象牙安魂住宅
252	2. New Design Languages for Old Styles 2. 用现代语言诠释传统模式
254	*Astor Apartment* 阿斯特公寓
260	3. Modern Design Concepts in Neo-Classical Homes 3. 新古典家居中的现代设计理念
262	*Down Street Apartment* 唐郡街公寓
270	4. Smart Homes in Neo-Classical Style 4. 智能新古典家居
272	*Mike & Sheila Mokhtare Family Residence* 莫克特别墅
282	5. Humanism and Sentiments in Home Design 5. 家居设计的人性化和情感化
284	*Meindorf* 梅恩道夫城堡
294	**Index** 索引

CHAPTER I
OVERVIEW OF NEO-CLASSICAL HOME DESIGN

新古典家居设计概述

1. Historic Development of Home Design

Home is the space that provides shelter and development opportunities for the human beings. It is an important venue of human life and has evolved from cave in ancient times to today's villa of intelligentised system. It has gained greater significance than a simple living space as it not only meets the need of "living", provides an ideal place for social activities, but serves as a carrier of human emotions and aesthetic interest.

1.1 Prehistoric Times

The science of home design is not officially established until the 1960s, yet the bahaviour can be traced back to a long time ago when decorative patterns of animals and plants appeared on cave walls as soon as people started living in natural caves. As human civilisation progresses, other forms of dwelling are created like houses built of branches and animal skin, Eskimo igloo (See Figure 1), and nomadic tents. But due to geographical conditions and construction techniques, these places of residence are not yet home design in a contemporary sense.

1.2 Ancient Egypt

Later in ancient Egypt, the creation of homes gained a certain degree of aesthetics. Although there was no complete relic of ancient Egypt residences for reference, we can still get a general idea of the situation then. Since there was a lack of wood for construction purposes, plus stone materials are mainly used in temples and tombs, common residences were made of adobe bricks which are made from a mixture of mud and straw. These square bricks are laid into walls as the roof is covered by straw mattress. There was a courtyard in the centre of the house as a public space of the family. Decoration of common households is simple with few pieces of furniture: mud-made bench and simple table, together with mats as bed. Living places of pharaohs and the rich, on the contrary, are quite exquisite. As can be seen in some murals of the time, noble people of ancient Egypt enjoy beautiful furniture of appropriate proportion decorated with floral patterns and legs resembling animal claws. Some of these high-quality pieces are even kept till today. A number of luxurious furniture were unearthed in the tomb of Tutankhamun in 1922. They are all made of rare materials, one chair of which is decorated with gold, silver, and gemstones. Other than luxurious

1. 家居设计的发展

家居空间是人类生存和发展的载体，是人类生产生活的重要场所。从最初用来挡风遮雨的山洞，到如今智能化的别墅，它的意义早已超越了一个简单住所的概念，它不仅满足了人们"住"的需要，也为人类的生活起居、社会交往提供了便利，同时也承载了人们的情感寄托和审美情趣。

1.1 史前时代的"家居"

作为一门专业的学科，家居空间的设计在20世纪60年代才开始正式确立，但对居住场所的装饰却已经有了很长的历史。早在人类开始学会利用天然洞穴的时候，在岩壁上就出现了用来装饰的动植物图案。随着人类文明的进步，陆续出现了用树枝、兽皮等搭建的房屋、因纽特人的冰屋（图1）、游牧民族的帐篷等居所的形式，但这些居住空间的设计都是由地理环境、建造技术等条件决定的，还算不上是现代意义上的家居设计。

1.2 古埃及时期

到了古埃及时期，家居的设计开始有了一定的美学意义。古埃及虽然没有留下完整的住宅遗址供现代人参考，但在一些遗留物中依然可以看出当时的状况。由于当地缺少能用作建筑用途的木材，石材

Figure 1. Igloo is the residence built with snow. The dome is a signature feature of igloos. As it is built with snow bricks in strict sequence, the finish is smooth and solid which turns out to be an effective shelter from the cold wind and can keep the warmth inside. It is the temporary housing method created by Eskimos in the northern part of North America against extreme weather conditions and a unique form of residence.

图1. 冰屋是一种由雪块构造而成的居所，最独特之处在于它的圆顶，是用一块块的雪块按规律堆砌而成，外观平整而且非常牢固，能有效地隔绝外面的寒风，也能保持室内的温度。冰屋是北美洲北部原住民因纽特人度过漫长寒冬的临时居所，是在极端天气下产生的一种特殊的住宅。

furniture, people of ancient Egypt also liked to decorate their homes with bright colours including red, green, and blue. Murals and other decorations on walls are also commonly used in this period.

As the glory of ancient Egyptian civilisation faded away, these exquisite designs withered. However, the art and design philosophy casts a profound influence on Mediterranean and European region.

1.3 Ancient Greece

Due to similar geographic conditions, the structural systems in ancient Greek and ancient Egypt residences are comparable. Houses are arranged around courtyard which accommodates rock garden, pavilion, fountain, and colonnade. The wall facing the street is window-less, often with no decoration. Bricks are used in some cases for interior flooring, while walls are painted or decorated with murals. As can be seen in some paintings of this time, home decoration of ancient Greeks follows a curvy elegant principle of exquisite setting.

Even though a considerable number of residences from ancient Greece are kept to this day, compared with temples and tombs, there are few houses and homes of great significance. Thus related study and research can only proceed on the basis of relics and paintings.

1.4 Ancient Rome

When it comes to ancient Rome, new changes took place in the form of residences. Other than following the courtyard house of ancient Greek style, there were apartment-style residences and vacation villas for the wealthy people. As the number of urban population increased, residences with large courtyards ran out of fashion, apartment-style housing prevailed. It was a brick and concrete structure of four to five levels with windows on each floor, quite similar to modern apartments. Ground floors of these structures were used for commercial purposes, and the upper floors were separated into individual apartments. Since these apartments are mostly for the poor, the interior was rarely furnished. Yet this form of residence cast great influence on the development of residences in a general sense, as it differentiates itself from traditional residence format and meets the growing need of housing of the time. It became a common housing form in later Roman Empire, yet the issues of limited space, vulnerability of fire, and proneness to collapse coexisted.

也只是用来修建神殿和陵墓，所以，普通住宅的建筑基本由土坯砖建成，这种土坯砖通常是泥巴和稻草的混合物，制作成方形，垒成墙壁，而房屋上面则是草顶覆盖。住宅中心带有一个庭院，是家庭中的活动中心。普通人的家居内饰很简单，很少有家具，一般只有泥制的长凳和简单的桌子，用席子当作床铺。而法老和富人们的居室则很考究，从一些壁画中可以看出当时贵族们的日常生活，家具的设计装饰丰富，外观优美，常饰有花卉图案，腿部雕刻成动物的爪形，并且尺寸合适，做工精湛，一些家具甚至保存至今。在1922年发掘的杜坦赫曼（Tutank Hamun）法老墓中，出现了许多豪华的家具，全部采用了珍贵的材料，其中一把座椅还镶嵌了金、银、宝石。除了装饰豪华的家具之外，古埃及人也喜欢用浓烈的色彩装点居室，明亮的红色、绿色、蓝色都很常见，墙壁上也常带有壁画和其他花样。

随着古埃及文明的衰落，这些精美的设计也随之消亡，但古埃及的艺术品味却对地中海地区以及欧洲地区都产生了影响。

1.3 古希腊时期

古希腊与埃及的地理条件很相似，因此住宅结构也很类似。住宅围绕着庭院布置，庭院中有假山、凉亭、喷泉等，周围带有柱廊。临近街道的一面不开窗，也没有装饰。房间内部有些是用砖来铺地，墙面也有粉刷，或带有壁画。从一些绘画作品中可以看到，古希腊家居空间内部陈设很精美，家具多用曲线造型，设计优美。

虽然住宅的数量很多，但相对于那些保存至今的神庙、陵墓来说，古希腊时期却并没有著名的住宅遗留下来，所以对此的研究也只能依靠一些遗址和绘画来猜测了。

1.4 古罗马时期

到了古罗马时期，住宅的形式有了新的变化，除了延续了古希腊时期的庭院式住宅之外，还出现了不同规模的公寓式住宅以及专门为富人休闲度假使用的别墅式住宅。随着城市人口的增长，占地面积较大的庭院式住宅已经不能满足需求，于是，出现了公寓式住宅这种形式。这种公寓式住宅与现代的公寓很类似，一般有四五层，由砖和混凝土建造，每一层都有窗户，底层用作商铺，上层分隔成住宅出租。这些公寓一般都是提供给穷人居住，因此内部陈设也很简单，但这种公寓的出现对住宅的发展有很大意义，它改变了住宅的传统模式，满足了当时日益增长的住房需求，在罗马帝国后期很普遍，但同时也存在着空间狭小、易发生火灾、易倒塌等问题。

奢侈的别墅是为富人们娱乐休闲所建，一般修建在山坡、海边等可

Luxurious villas were built to serve recreational purposes of the well-off. They were normally constructed on spots of great scenery like sloped hills or seaside covering a relatively large area with components of porch, yard, and big garden. Other than dining rooms and bedrooms, there were also less functional spaces of bathrooms, small library, etc. The interior followed a glamorous style as well while the rooms were decorated with various themes. Murals, mosaic decoration, and even tapestries could be found on walls. Furniture was mostly made of stone, bronze and wood with ornamental metal or ivory, created in ancient Greek style. (See Figure 2)

In addition to residential forms, advanced techniques like water supply and drainage systems flourished in ancient Roman period. They even created a unique heating system. These techniques are of considerable significance on residence design, some of which still have their influence on contemporary home design today.

1.5 The Middle Ages

After the Roman Empire collapsed, arrived the long, dark Middle Ages of Europe. A pyramid social structure came into form after the feudal hierarchy established. On the top of the pyramid were kings and the noble, the bottom farmers and slaves, knights and those with titles were in between. The farmers of lower level were controlled and exploited by the upper level via land ownership and tax payment, which leads to constant uprising. There were also frequent wars and conflicts among the upper level. Residences of this time were created on this chaotic social background.

In early Middle Ages, social chaos and economy recession prevailed. Farmers dwelled in humble places of mud or wood roofed with straw, often with only one room. Light hardly got to the inside and there was little furniture. In the colder area in the north, some houses were equipped with fireplace. As for the houses of land owners, warehouse, mill, wine cellar were included in addition to common bedrooms. There was courtyard and surrounding walls. Benches, tables, ceramic cups and jars could be found inside the house. Brick or wood flooring was available as well. (See Figure 3) Nobility and the upper society lived in small castles in the early days while later in the late Medieval some started to build mansions instead of castles. These were large houses with plenty of space and finer decoration. Bright-coloured fabrics were used on tables and benches. Glass was also available at this time which allows better light penetration.

以欣赏风景的地方，面积较大，带有门廊、庭院、大花园，除了饭厅、卧室等功能区之外，还有浴池、小型图书馆等。别墅的内饰一般也很豪华，不同的房间会有不同的装饰主题，墙壁上有各种题材的壁画，或者马赛克的装饰，也会有挂毯之类的点缀。家具形式由古希腊延续而来，由石头、青铜、木材等制成，还会镶嵌珍贵的金属、象牙等装饰品（图2）。

除了住宅的形式之外，古罗马时期的一些技术也很精湛，他们有完善的供水系统和排污系统，甚至还有一套独特的供暖系统，这些技术对住宅的发展有着至关重要的意义，有些技术对现代家居设计依然产生着影响。

1.5 中世纪

在古罗马政权瓦解之后，欧洲地区迎来了漫长而又黑暗的中世纪，封建等级制度的形成使社会形成了金字塔形的结构，从顶层的国王、贵族，到有封号的阶层、骑士，最底层是农民和农奴，上层凭借对土地的拥有权和纳税对下层农民剥削控制，导致农民起义不断。除此之外，上层之间也战争频繁，混乱的社会模式是当时住宅形式的基础。

中世纪早期，社会混乱分裂、经济衰退，下层农民的住宅很简陋，泥土或木头建成的小屋一般只有一间房，房屋坡顶上覆盖着草，室内光线很暗，几乎没有什么家具，北方地区住宅有时会带有壁炉取

Figure 2. An ancient Roman residence in historic painting
图2. 绘画中的一处古罗马住宅

Medieval houses emphasised overall practicability and functional interior design. These features made them extremely sturdy that some houses are still in use today in the way they were first built.

1.6 The Renaissance

From the 15th century, new thoughts started to emerge in Europe and new trends prevailed in architecture and interior design. The latter was greatly influenced by Roman style, emphasising proportion and symmetry. Residences were more of a space for social activities than solid castles while multiple changes took place in home design on the basis of former design concepts leading to a larger variety and finer details. Other than the conventional tables, chairs and beds, there were also bookcases and wardrobes in Renaissant homes. More complex designs were created in each furniture category, like folding chair, according to specific functions. The development of art and craft provided possibilities for diversified furniture, as in wealthy households of this period, wardrobes, large cabinets and sideboards with carved patterns seemed quite common. Apart from the changes in furniture, fabrics became luxuriant as well, among which silk and velvet materials were most popular in the early period. Used on pillows, benches and cushions, these bright fabrics add to the luxurious ambiance of the space.

As society developed and wealth accumulated, home design witnessed the transition from simple functionality to magnificent aesthetics. Thus was the main feature of Renaissance.

1.7 Baroque Ages

In the 16th and 17th century, people in Europe enjoyed greater freedom of thoughts and higher living standard. There were more scientific inventions and the art of Baroque was created. This form of art inherited the emotional focus of the Renaissance, pursuing a dynamic form and free expression. Curve lines and oval pattern were preferred in decoration with a strong colour palette. (See Figure 4) As monarchy and religious reformation were taking place in Europe then, all art was supposed to serve the King and the Pope. Architecture and interior decoration of Baroque style were more commonly used in churches and palaces, not normal households. A distinguishing feature of Baroque style is the degree of complexity and exquisiteness as not only the ornamental pattern of leaves, shells and swirls appeared on walls and ceilings, but marble

Figure 3. Restored residence of medieval middle-class displayed in a museum
图3. 博物馆中展示的中世纪中产阶级住宅复原场景

暖。而拥有土地较多的庄园主住宅则更大，除了自己居住的卧室之外，还包括一些仓库、磨房、酒窖等，有庭院，周围围着院墙。室内有简单的长凳、桌子，以及陶制的杯子、瓦罐等。地面铺有砖或地板（图3）。早期的贵族和上层社会一般居住在小型城堡中，到了中世纪晚期，一些人放弃了城堡开始修建大住宅。这种住宅宽敞舒适，室内装饰也精美了许多，开始有了挂毯一类的织物饰品，长凳和桌子上也有颜色明亮的覆盖物，玻璃的运用也越来越多，窗户尺度的增大为室内增加了采光面积，这些都成为了中世纪晚期住宅的重要特征。

中世纪的住宅建筑实用性强，室内设计也更强调功能性，这些特点使它异常坚固，有些住宅甚至直到现在依然发挥着当初的作用。

1.6 文艺复兴时期

从15世纪开始，新的思想开始在欧洲地区出现，与此同时建筑和室内设计也被新的风格所主导。室内设计受到古罗马风格的影响，强调比例和对称。住宅不再像中世纪一样是坚固的堡垒，而是有了更多用于社交的空间。家居设计在吸收了以往设计精华的基础上有许多变化，其中最大的变化就是室内的家具品种更多，也更加精致。文艺复兴时期的家居空间内除了之前的桌、椅、床之外，还有了书柜、衣柜等形式，每种家具也根据用途有了更复杂的设计，例如便捷的折叠椅。手工艺的发展为家具的多样性提供了更多可能，在富裕的家庭里，雕花的衣柜、大型的储物柜、餐具柜都很常见，并且这些家具都与装饰工艺结合在一起，以往用在建筑细部的莨苕

Figure 4. The Paris Opera House is famous for its luxurious Baroque interior design
图 4. 巴黎歌剧院，以豪华的巴洛克室内风格著称。

flooring of more colours including red, yellow and gold were applied. Gypsum plaster and mirrors were used on walls for a glorious effect. Furniture took on a more exquisite design with curvy outline and images of plants or human figures, sometimes with precious ivory, gold or silver inlay. The legs were carved into the bulbs. Luxuriant fabrics were also used in some chairs. (See Figure 5) The interior of Baroque style is of masculine quality, emphasising exaggeration, tension and irrationality. Though it is unique in so many ways, the limitation to churches, palaces and mansions leaves the houses of ordinary people in Renaissance with only a hint of Baroque.

1.8 Rococo Period

Rococo style originated in France at the beginning of the 18th century, when the king of France Louis XV moved from the Versailles Palace to the delicate new palace in Paris. This comfortable and relaxed decoration style gained popularity first in the upper class

叶、天使、蔓藤等图案也被用在了家具的设计中成为表面的装饰。除了家具之外，织物也华丽了很多，丝织品和天鹅绒在早期最为流行，色彩明亮的织物覆盖在枕头、长凳和垫子上，为家居增添了奢华的味道。

随着社会文化的繁荣和社会财富的积累，家居空间的设计开始从简单实用到复杂华丽转变，文艺复兴时期正是承担着这种转变的过渡期。

1.7 巴洛克时期

16世纪到17世纪，欧洲地区思想更加自由，物质也更为丰富，与此同时产生了了许多新的发明。在这样的背景下，产生了巴洛克艺术。这种艺术风格继承了文艺复兴时期艺术对情感的注重，追求动态的、自由的形式，色彩强烈，富于装饰性，常喜欢用曲线和椭圆形图案（图4）。由于当时的欧洲正处在君主专制和宗教改革的时代，所以艺术都是为君主和教皇服务的。巴洛克的建筑和室内装饰

and the nobles, then among the bourgeoisie. Since stability and comfort prevailed, these people enjoyed an active social life. For this reason, the ground floors of the residences were usually grand luxurious sitting room. There would be a display of the owner's art and painting collection other than exquisite decorative details. The first floors were often bedrooms, while the higher levels were used for storage. Residences of Rococo style emphasise well proportioned space, continuous wall-and-ceiling ornaments, and diversified decorative elements. Rural scenery and landscape, or patterns of swirled shells were common decorative themes. The design of furniture employed curved outline and gorgeous colour palette, emphasising surface decoration via means of gold plating, colouring, inlay, and carving. Similar to Baroque style, Rococo decoration was for the noble and aristocrats. Its luxurious and complicated features were favoured by the upper class of France, which also reflected the uprise of feminism. (See Figure 6)

1.9 Neo-Classical Period

During the late 18th and early 19th century, Europe witnessed dramatic changes: the French Revolution, the Industrial Revolution, the rise of the bourgeoisie, and the accumulation of social capital, resulting in a rebellion to the art of Baroque and Rococo. The excavation of Roman relics provided a basis for reviving Classical art after a rigorous and elegant aesthetics. This new art form was guided by the revival of Classic art from ancient Greek and Roman Empire, advocating the classical design principles of symmetry, proportion and order yet surpassing Classic art as it was more delicately beautiful.

Neo-classical home decoration adopted different forms in European and American countries. In the early stage of French Neo-classicism, home design opted for Classical elements like straightforward outline and decorative patterns of the Tricolour, sword and spear related to the French Revolution. When it comes to the Napoleon period, the theme of war and Napoleon's initial also appeared in residences. The colours of red and black were popular in this time, while furniture, mirrors and fabrics with gold-plating were signature items of the period. In Great Britain, houses were built around squares. They usually came in four to five levels, facing squares or streets, the basement of which was used as kitchen, ground floor as sitting room and dining room, first floor as entertainment room, second floor as master bedroom, third floor as guest room and children's room, and

艺术也更多地表现在教堂和宫殿中，普通人的住宅则很少被关注。但是在富人和贵族们的宫殿中，我们依然可以领略到当时家居设计的一些蛛丝马迹。复杂和精美是巴洛克风格区别于以往最显著的特点。不仅在墙面和顶棚会有树叶、贝壳、涡卷等雕刻，地面的大理石也会有丰富的色彩，如红、黄、镀金等。墙面有时还会用石膏或镜子装饰，营造出华丽又迷离的氛围。家具的设计更为精美，柜类家具表面都会有曲线装饰，并带有植物、人像的图案，有时还会镶嵌珍贵的象牙、金或银，腿部雕刻成球形，体积巨大，而椅类等家具还会带有华丽的纺织品（图5）。巴洛克的室内空间强调夸张、力量和非理性，是带有男性气质的装饰风格，虽然它的特点很突出，但主要都是存在于教堂、宫殿和豪宅中，普通人的住宅还停留在文艺复兴时期，只是偶尔会有一些雕刻或曲线的装饰显示出一点巴洛克的痕迹。

Figure 5. Hand-carved wooden dining chair in Baroque style
图 5. 巴洛克风格实木手工雕刻餐椅

the top floor as the servants' room. Gypsum plaster was the common option for ceiling decoration. There were sculptural fireplaces in the main rooms. And decorative details of the interior could be simple or luxurious according to the owner's social position, yet oil painting or mirrors were essential whatever the case. (See Figure 7) Neo-classical home design in Britain featuring logical consistency and rigorous detail treatment is highly regarded in the design history. Outside Europe, Neo-classical style also cast a profound influence in America. Since the first settlers stepped on the American continent, local residence design has always followed the trends in Europe and it changed from a simple, natural style to delicately beautiful in the 18th century. The module of local residences resembled that

1.8 洛可可时期

洛可可风格出现于18世纪初的法国，当时的法国由国王路易十五统治，他将宫殿从凡尔赛宫迁到了巴黎城中，新的宫殿精致而小巧。与此同时，社会上也开始流行这种舒适轻松的装饰风格，首先是上流社会和贵族们的竞相模仿，然后资产阶级也开始纷纷效仿。这时的上流社会生活舒适安逸，喜爱社交，因此，在住宅的底层都会有豪华的客厅，这种客厅带有沙龙的性质，除了奢华的装饰外，还会展示主人收藏的艺术品、绘画等。二层主要是卧室，而更高层一般会用来当做储藏室。洛可可风格的家居空间讲究和谐的比例、墙面和天花板装饰的融合以及装饰的多样性。装饰的题材多为田园和风景，或是涡卷贝壳纹样，造型以曲线为主，色彩艳丽柔和，家

Figure 6. The Wies Church in Germany is a typical Rococo architecture built by renowned architect Dominikus Zimmermann. At the centre of the building is an oval atrium 29 metres long and 25 metres wide, with the surrounding columns supporting the colourful dome. A vacant throne and the closed gate to heaven are painted on the dome.

图6. 德国威斯教堂（Wies Church）是洛可可风格的典型代表，由著名建筑师米尼库斯·齐默尔曼（Dominikus Zimmermann）设计，教堂的主体是长29米、宽25米的椭圆形中殿，殿中环绕的圆柱支撑着色彩斑斓的穹顶。穹顶一端绘着虚位以待的王位宝座，一端绘着紧闭大门的天堂，整个教堂充满了缤纷的色彩。

Figure 7. Symmetrical furniture, white marble fireplace and large area of red colour make up the Neo-classical style with a hint of British heritage.
图7. 对称的家具摆放、白色的大理石壁炉以及大面积的红色调构成了有英国特点的新古典风格。

of Britain, but with wooden structure and symmetrical layout. The ground floor accommodated reception hall and dining room, the first floor was used for bedrooms, while kitchen and storage rooms were arranged on both sides of the building. Timber inlay was employed in the interior wall and the design of ornamental fireplace was just same as in British homes. Later the design pattern took on more French features with linear furniture, decorative inlay, copper handles, animal-shaped legs, and all in all, bigger volume and more weight.

1.10 Modern Homes

From the beginning of the 19th century, a variety of home design approaches flourished: the art and craft movement, New art movement, Eclecticism... Yet these trends were all somehow

具设计强调表面装饰，多用镀金、着色、镶嵌、雕刻等各种手法相结合，追求华丽轻巧，并且采用不对称的设计，以表现浪漫飘逸的色彩。与巴洛克艺术相同，洛可可风格也仅限于宫廷和贵族们的府邸，华丽繁琐的装饰迎合了上层社会的爱好，也反映了当时女权主义呼声的高涨（图6）。

1.9 新古典时期

18世纪末19世纪初，欧洲社会发生了很大变化，法国大革命、英国的工业革命、以及资产阶级的兴起和社会资本的积累导致艺术领域产生了对巴洛克和洛可可风格的反叛，而古罗马遗址的发掘又为复兴古典艺术找到了根据，追求真实严谨、典雅优美的新古典风格应运而生。这种风格以振兴古希腊古罗马艺术为宗旨，继承了古典主义对称、比例均衡、秩序严谨等设计原则，但又比古典主义更精美。

related to the Classical style. The brand new home design concept of Modernism did not appear until the 20th century. It touched every aspect of art practice, representatives in architectural and interior design being German architect Walter Gropius, Ludwig Mies van der Rohe, French architect Le Corbusier and American architect Frank Lloyd Wright. They brought up the concept of "form follows function" meaning that modern homes should be designed on the basis of functionality, taking into consideration the elements of human needs. It focused on the analysis of the interior and exterior, understanding specific needs of each family member and all in all, the division and integration of space. In addition, technology development and advanced materials provided the sources for diversified modern homes. Electric lighting system, modern kitchen and bathroom facilities, various new materials all contribute to a more comfortable residence. (See Figure 8) Today, the evolution of home decoration goes on, be it Post-modernism of passion and fantasy, Deconstruction in restructuring confusion, or the nostalgic revival of Classic aesthetics. Currently, green and intelligent homes are the key words of future development yet no one knows for sure what our homes will look like in a few years' time. One thing to be certain is that whatever the case, human beings are always working on the creation of a better residential module and the exploration never stops.

2. Key Points in Neo-Classical Home Design

A successful residence design needs to cover and reflect the information of a family's lifestyle and pursuit. Nowadays, this means demonstrating certain aesthetic appeal, other than possessing basic functions. As an elegant and retro art form still popular in Europe and the United States, Neo-classical decorative style stands for good taste and high social status. A major issue in the application of Neo-classical style is to meet both needs of functionality and form, making it comfortable and at the same time environmentally friendly.

2.1 Living Room

Living room is the public space in a house for visitor reception, leisure and entertainment. It takes up a relatively large area and is mostly frequently used, which makes it a focal point of design. Living rooms accommodate a considerate number of elements and require a lot of continuous space, especially in Neo-classical households. Therefore houses of multiple levels often take full advantage of the elevated

新古典主义家居的设计在欧洲及美洲各国家有不同的表现形式。法国新古典主义早期，家居的设计更倾向于古典主义，采用了更多的直线条，并且在装饰上有很强的法国大革命痕迹，常用三色旗、剑和矛做装饰的主题。而到了拿破仑时期，除了这些题材之外，还加入了战争题材，以及拿破仑名字的首字母"N"。红色和黑色是这一时期的流行色，在工艺上常用镀金，镀金的家具、镀金的镜子以及镶着金边的织物都是这个时期显著的标志。而在英国，城市住宅一般都围绕着广场建造，住宅面对着广场和街道，有四五层高，地下室用作厨房或仓库，一层是客厅和餐厅，二层是娱乐活动室，三层是主人的卧室，四层是客房和儿童房，顶层一般供佣人居住。室内都会带有石膏装饰的顶棚，主要的房间也会有精雕细琢的壁炉，室内陈设和家具根据主人的地位或简朴或奢华，但都会有油画或镜子等装饰（图7）。英国的新古典风格家居设计有逻辑上的一致性，细部处理也很严谨，在设计史上很受推崇。除了欧洲之外，新古典

Figure 8. Modern home design features elegant approach and material diversity, with humanised design as the first priority.
图8. 现代家居设计简洁大方，材料多样，以人性化为首要目标。

Figure 9. A medium to small living room should be centred around the fireplace with symmetrical chairs. Tea table should employ a symmetrical form as well. Simple sofa for two is chosen in order to provide access. A flexible area is saved at the end of the living room for piano.

图9. 面积并不大的客厅以壁炉为中心，摆放了对称的椅子，茶几也是对称的形状。为了防止过道的拥挤，采用了简单的双人沙发，在客厅一端还专门开辟了一定的面积摆放钢琴。

upper part to create better view and introduce more daylight. Together with large windows, heavy textural drapes are an essential of Neo-classical style. Besides, the layout of living rooms follows a consistent and rational pattern. Roman style columns can be used in spacious living rooms for ornamental and divisional purposes. Order is a major principle, as inherited from the ancient Roman style and ancient Greek houses: columns and furniture need to be presented in a symmetrical way, with the fireplace as the centrepiece surrounded by sofas in a U to create an accessible aisle. Flexibility is another issue to consider since living rooms sometimes hold reading and audiovisual activities. Durable marble flooring of slip-resistant features is a preferred option, often used with carpet. Since living room is the space the owners meet and entertain their guests, it is supposed to cater for both personal and general aesthetics. (See Figure 9)

2.2 Dining Room

Dining room is the place for everyday meals and treating guests, usually next to kitchens especially in open kitchens in Europe and the States. In some cases, it is located near to the living room. The design of dining room needs to be consistent with that of the sitting

风格在美洲同样有很大影响。自从第一批殖民者踏入美洲大陆后，这里的住宅形式就一直追随着欧洲的脚步，到了18世纪，美洲地区的住宅风格也从原本的简单朴实转向了华丽精致。这里的城市住宅形式与英国住宅很相似，但更多是由木材建造，室内空间采用对称的布局，一层是用来会客的大厅和餐厅，二层是卧室，厨房、储藏间等安排在住宅的左右两侧。室内墙面用木板镶面，装饰性的壁炉与英国住宅如出一辙。而到了晚期更倾向于法国的设计，家具多采用直线条，有镶嵌物、铜制拉手、爪形腿，但更为沉重，体积也较大。

1.10 现代家居

自19世纪开始，家居设计的风格层出不穷，工艺美术运动、新艺术运动、折中主义……但这些风格依然离不开古典的范围。直到20世纪，现代主义的出现为家居设计带来了全新的形式。现代主义涉及到艺术的各个方面，在建筑和室内设计领域的代表人物主要有德国建筑师瓦尔特·格罗皮乌斯（Walter Gropius）、路德维希·密斯·凡德罗（Ludwig Mies van der Rohe），法国建筑师勒·柯布西耶（Le Corbusier）和美国建筑师弗兰克·劳埃德·赖特（Frank Lloyd Wright）。他们提出了"功能决定形式"的主张，认为现代家居要以功能为出发点，充分考虑人的因素、满足人的需求，同时分析室内环境和室外环境，了解家庭成员中的各自需求，对空间进行

room, centred around the dining table with chairs and cupboards. It requires good lighting and ventilation conditions, and therefore decorative elements need to give way to free access. Furniture of Louis XV and XVI style is desired for a complete Neo-classical ambiance. If there is enough space, an ornamental fireplace can be added in the centre of the room. Windows in the dining room should be as large as possible and thick curtains should be avoided. Lighting for dining room should be soft and therefore dazzling crystal lamps are not recommended. (See Figure 10) Also, layout and setting in dining rooms ought to be simple and flexible to cater for daily dining and guest entertaining.

划分和整合。另外，随着工业技术的发展和新型材料的产生，家居设计也变得更加多元化，全新的电力照明系统，现代的厨房和浴室设备，层出不穷的新型材料让家居空间变得越来越舒适，也逐渐地改变着人们的生活方式（图8）。而在当代，家居空间的变化仍未停止，无论是充满激情和幻想的后现代主义，还是在混乱中重组的解构主义，亦或是力求再一次回到古典的传统复兴，他们都在为建立更美好的居家环境而进行着探索。如今，绿色和智能是新时代家居的关键，而未来的家居空间会是何种形式还是未知，但可以肯定的是，无论即将出现的家居空间是什么样子，人类都会一如既往地为创造更适宜的居住场所而努力，对家居设计的探索也永远不会停止。

Figure 10. Dining rooms need to be well-lighted and comfortable. This dining room has big windows, neat furniture and exquisite plate decoration.
图10. 餐厅需要明亮温馨的氛围，这间餐厅有宽大的窗户，餐桌和餐椅简洁古朴，还用了瓷盘做装饰。

Figure 11. The colour continuity in the dining room and kitchen helps establish a coherent effect.
图11. 与厨房相连的小餐厅在色彩上与厨房有一定的连续性，保持了连贯的风格。

In cases where dining room is next to the kitchen, a consistent design should be employed. (See Figure 11) Generally speaking, kitchens should be tidy and orderly, well ventilated. Though there is no place for excessive decoration in kitchens, the choices of cupboards, flooring pattern and material can still reflect the concept of Neo-classical art. Wooden cupboard of white lacquer is a common choice while those of natural colour used with stone flooring in similar shades provide an ideal Neo-classical effect. Similarly, too much decoration is not necessary in kitchen design. Symmetrical layout and furniture arrangement are the better option for establishing Neo-classical style.

2.新古典家居的设计要点

家居空间是根据家庭的需求、生活方式、精神境界来设计的，发展到今天，除了满足功能性要求之外，更需要有一定的审美情趣。新古典风格以复兴古典为宗旨，典雅大气，在众多设计风格中已独树一帜，在欧美国家乃至其他地区依然是品味和地位的象征。然而在家居设计越来越讲究舒适和环保的今天，如何让这种风格同时满足功能和形式的双重要求，是家居设计中的关键。

2.1 客厅设计

客厅是用于会客、休闲、娱乐的公共空间。在住宅中的利用率很高，所占面积也较大，是设计中的重点。客厅的设计需要开阔的空间，尤其是新古典风格，运用的元素很多，如果过于狭窄会造成拥挤，失去了新古典风格原本的含义。因此，一些多层的住宅会充分利用上层部分形成挑高的空间，使视线更开阔，也有更好的采光效果。大面积的窗户也必不可少，材质厚重的落地窗帘是新古典风格的重要特点。另外，客厅空间的划分需要有一定的连贯性和合理性。如果空间宽敞可以使用罗马柱，既可以作为装饰也能起到分割的作用。新古典风格秉承古罗马和古希腊时代的秩序，无论是柱式还是家具的摆放都需要用对称的方式，一般以壁炉作为中心，周围的沙发以U形环绕，但是要保证中间过道的顺畅。客厅的使用频繁，有时还会承担阅读、视听等功能，所以还应该留有一定的弹性空间，但在风格上要与整体一致，材质的选择上需要耐磨并且防滑，大理石材质硬朗，气质华贵，是新古典风格的首选，再搭配大面积的地毯会起到事半功倍的作用。客厅承担着会客的功能，所以在风格的表现上更要兼顾个性化和大众的审美观念，让更多的人易于接受，同时也不能为了表现风格使用过多装饰而忽视舒适性（图9）。

2.2 餐厅设计

餐厅是家庭用餐、宴请客人的地方，一般会与厨房连在一起，尤其是欧美国家的开放式厨房，有时也会靠近客厅。餐厅的设计要与客厅保持一致，布局一般以餐桌为中心，周围是餐椅，有时还会有储物柜。餐厅需要良好的照明、通风条件，所以在风格的塑造上可以适当简化，留出足够的空间供人的活动。在餐桌和餐椅上选择路易十五或路易十六风格都可以达到很好的效果，再搭配同样类型的储物柜，就可以完整地塑造新古典的风格。如果有足够的空间，还可以在餐厅中心安排装饰性的壁炉。窗户可以尽量宽大，窗帘的选择不宜过于厚重，防止积累灰尘。餐厅的光线要以柔和温馨为主，不宜选用过于耀眼的水晶灯（图10）。另外，餐厅的布局还应该具有一定的灵活性，既能满足日常家庭成员的用餐需求，也可以满足宴请的需要，所以餐厅中的陈设不应过多，也不应过于复杂。

Figure 12. The bed is the core of a sleeping area and all furniture including bedside tables and closet needs to be consistent. (Rendering)
Figure 13. This bedroom is relatively large, with independent audio-visual zone and dressing area and the audio-visual zone is separated by pillar elements. (Rendering)
Figure 14. For a smaller bathroom, a simple modern style is preferable. In this large bathroom, however, pilasters and crystal lamps are used to create a consistent Neo-classical style. (Rendering)

图12. 睡眠区中床是中心，四周的家具，床头柜、衣柜需要与床的风格一致（效果图）。
图13. 这间卧室面积较大，有单独的视听区和梳妆区，在视听区用了柱式做分割（效果图）。
图14. 对于小面积浴室来说，一般会采用简单的现代风格，而在这间大面积的浴室中却用了壁柱、水晶灯等元素创造出了与卧室一致的新古典风格（效果图）。

2.3 Bedroom

Bedrooms serve the purposes of sleeping, resting, storage and dressing up, and therefore they are located far from the entrance, at the inner part of the house. The level of privacy and comfort is the major issues to consider in the design process of bedrooms to ensure satisfying functionality. Sleeping area, as the core of a bedroom, is often far from the bedroom door. The bed and bedside form the backbone of the sleeping area and both their position and number are decided according to the symmetrical principle of Neo-classical art. Traditional four-poster bed adds to the romantic Classical features of a bedroom, yet it is too complicated for modern homes, and therefore it is used without the canopy and only the posters. A footstool at the end of the bed is another characteristic item. The storage function is mainly fulfilled by wardrobes that match the bedside tables. Dressing area sometimes appears in the bathroom, and in common households, bathrooms are more functional rather than presenting specific decorative style. (See Figure 12, 13, 14) Due to the basic feature of privacy, bedroom design seeks peace and comfort. Wood flooring and carpet, extravagant wallpaper and wall paint of smoothing colours are all favourable choices to create an elegant Neo-classical space. It also varies according to the users' needs, i.e. degree of floor evenness and slippery, comfortable level of furniture.

2.4 Basic Spatial Units

Basic spatial units include entrance, corridor, storage room, guest bathroom, etc. These are auxiliary areas whose design could affect

如果餐厅与厨房相连，厨房的设计与餐厅应该一致（图11）。一般来说，厨房的设计需要整洁有序，并保持通风良好。厨房中虽然没有需要过度装饰的地方，但在橱柜的选择、地面的铺设等处依然可以体现出浓厚的新古典风格。白色的木制橱柜是常见的选择，另外，原木色的橱柜和同色系的石材地板也可以很好地体现新古典的氛围，又更易于清洁。同样，厨房的设计也不宜有过多装饰，通过对称的布局、界面的设计以及家具的摆放来体现新古典风格是更好的选择。

2.3 卧室设计

卧室的功能主要有睡眠、休息、储藏、梳妆，一般都会安排在远离入口、住宅最里面的位置。卧室空间讲究私密性、舒适性，所以在设计中需要更多地关注细节的处理，以保证卧室的实用功能。睡眠区是卧室的核心，一般会在远离门口并且相对稳定的方向，床和床头柜组成了基本的睡眠区，新古典的对称原则决定了床的中心位置和左右两侧相同的床头柜。传统的四柱床可以增添卧室的古典色彩，但在现代的家居中，这种床过于复杂，有时会取消四周的帷幔，只用四角的四根柱子代替，或者直接使用现代类型的床。另外，床尾放置一个矮凳也是新古典风格重要的特点。储藏区以衣柜为主，与床头柜同系列的衣柜必不可少，形式同样需要遵循对称的原则。梳妆区有时会被直接纳入卫生间中，在一般的家居设计中，卫生间一般不会特别强调风格的呈现，而是以现代材料呈现的实用性为主（图12、13、14）。由于卧室的私密性特点，其界面的设计应该尽量温馨、宁静。地面以木制地板和地毯为主，墙面上装饰质感华贵的壁纸、墙布或是色彩淡雅的涂料都很适合。新古典风格典雅大气，因此需要较大并且规则的空间来实现，在家具的摆放上也不宜过于拥挤。除此之外，根据居住者的不同，卧室的设计还会有

the overall ambiance of a house. An entrance is the transitional space between the inside and outside. It is the buffer and protection before the interior. Though there is no need for a lot of space at the entrance, it should be enough for clothes changing. A simple Louis XVI cabinet with antique craftworks or candlesticks together with a wooden-frame mirror, makes a perfect Neo-classical entrance. In terms of lighting, balanced wall lamps can be added to complement the main light source.

Corridors and staircases are circulation spaces of great significance;

不同的要求，例如儿童的卧室或是老人的卧室，在保证基本的功能之外还需要额外保证地面的平整程度、防滑程度，家具的舒适程度等。

2.4 基本空间设计

基本空间包括玄关、走廊、储藏室、客用卫生间等。这些空间主要起辅助的功能，但如果设计不当也会影响家居的整体。玄关是连接室内外的过渡空间，在总体布局上应该作为进入室内的缓冲，使人不能对室内的情况一览无余。虽然玄关的空间不需要太大，但也应该有足够的更换衣帽的位置，一个简单的路易十六风格柜子，上面摆放复古的工艺品或烛台，再搭配木制框的镜子，就可以营造出有新古典风格的玄关。在光线上，玄关处也不宜过暗，除了主要的光源之外，也可以在墙面上增加对称的壁灯。

走廊楼梯间等处是必要的交通空间，也应该根据整体风格做必要设计，否则会过于沉闷，与家居整体风格不符。这些位置一般空间狭小，可以根据情况在墙面装饰油画、挂毯等，或者只是用镶板等做出一些规律的几何造型，产生一定的韵律感和节奏感。如果空间充

Figure 15. Wallpaper of vertical lines is used in the corridor to add depth to the space so it still feels spacious with cabinets and chairs in the way.
Figure 16. Symmetrical pilasters for corridor are a common decorative approach in Neo-classical decoration; the balanced use of chairs can avoid the monotony in long corridors.
图 15. 走廊用竖线条的壁纸使空间显得更有纵深感，宽敞的走廊即使摆放了柜子和椅子也不觉得拥挤。
图 16. 走廊墙面的对称壁柱装饰是新古典风格常用的手法，搭配对称的椅子避免了长走廊的单调。

thus their design should correspond to the style of the rest of the house. Paintings, murals or geometric patterns of decorative panels are ideal options for this area. Space allowing, a Neo-classical chair is also a good choice. Soft clean lighting is preferred, normally achieved by wall lamps. (See Figure 15) Stair armrest with delicate carving and quality carpet are usually used in Neo-classical staircases, (See Figure 16) especially in large houses with wide long stairs, which could greatly affect the overall atmosphere of the whole space.

3. Design Principles in Neo-Classical Homes

3.1 Colour Design in Neo-Classical Homes

Colour is the strongest and most expressive factor of an interior. A balanced colour palette provides the aesthetic basis for house styling

足，摆放一张古典风格的椅子也是很好的选择，另外，光线柔和的壁灯也不可缺少，但光线不宜过于绚丽（图15）。而楼梯这样的地方也需要精心雕刻的扶手、柔软的地毯为其增色（图16），尤其大型的住宅，楼梯的面积大，往往会在很大程度上影响室内空间的氛围，新古典风格常用弯曲的楼梯造型，再加上雕刻的铁艺扶手以及印花地毯装点。

3. 新古典家居的装饰原则

3.1 新古典家居的色彩运用原则

色彩是室内空间因素中给人感觉最强烈最直接的要素，与造型有着同等重要的作用。合理的色彩搭配体现了家居的美学基础，也能优化家居的形式美。

Figure 17. Adding lively colours in the white space helps attract people's attention to the red sofa before the fireplace.
图17. 以白色为主色调的空间加入若干跳跃的色彩，使人很容易把视线集中到壁炉前的红色沙发上。

and also optimises the formal beauty of the residence.

The colouring of an interior emphasises functionality as much as a harmonious atmosphere. Formal aesthetics comes first, and then is the unity with other decorative elements. There is a range of common Neo-classical colours: white, gold, grey, light blue, dark green, and dark red. Soft colours of lower saturation degree and much brightness are also used sometimes in a limited extent. Netural colour palette works best for normal households, but this doesn't mean that Neo-classical houses appear dim and pale. A clever contrast of colours, a hint of gold or floral patterns of multiple colours can all contribute to forming the Neo-classical highlight. (See Figure 17)

White and gold are a common colour combination in Neo-classical

家居的色彩搭配要在强化功能性的同时使室内的环境更具协调性和美感，首先要符合形式美，其次也要与环境和谐，并且运用合理、舒适，而新古典风格也有自己的搭配原则。新古典风格常用颜色很多，白色、金色、灰色、淡蓝色、深绿色、暗红色，以及其他饱和度不高并且较柔和的颜色，偶尔也会在小范围使用亮度较高的颜色。对一般的家居环境来说，中性的色调最为合适，但这并不意味着新古典风格的家居会很昏暗和苍白，色彩的碰撞对比、金色的加入、多种色彩的花卉图案都能形成新古典风格的亮点。并且只要避开过于耀眼的亮色系和不和谐的组合，任何色彩都可以合理地使用（图17）。

白色和金色是新古典家居中最常用的组合，在18、19世纪的城堡中，这种颜色搭配很常见。白色的镶板墙面饰以金边，加上金色为主的椅子和家具，显得愈加富丽堂皇。而在普通的住宅中，墙面的镶板可以用更简单的涂料代替，用少量的金色镶边，家具则可以选择暗红等其他颜色。白色和淡蓝色也是新古典风格中很好的搭配，白色和淡蓝色组合在一起清新淡雅，能创造出明亮又温馨的家庭氛围，组合中可以以淡蓝色墙面为主，搭配白色的天棚，床品，衣柜，为了防止单调，还可以加入印花的织物调和。如果空间需要更庄重典雅，可以选择暗红色和金色的搭配（图18）。暗红色自拿破仑时期开始就成为了新古典风格中的主要色调，在法国的枫丹白露宫中可以看到这种搭配的典范。在普通住宅中，如果觉得单纯的两种颜色过于富丽，可以减少暗红色的比例，如果是软装饰选择了暗红色，则可以加入一定面积的米黄或白色墙面。当然，无论哪种颜色的搭配，新古典风格的原则是统一和谐，同时也要满足住宅各功能区的要求，为住宅内生活的人带来愉悦。

3.2 新古典家居的界面设计原则

住宅内的空间是由水平空间和垂直空间围合而成，界面的大小和形状影响着室内空间的大小，界面的造型等效果也影响着室内的环境。不同位置和不同功能区的界面要求不尽相同，但都有一些基本原则，例如阻燃性、保暖性、隔音性、耐久性，以及美观的要求，另外也要有一致性，以免影响整体的协调。在玄关、客厅、餐厅这些敞开式的公共空间，地面的设计应该尽量统一，以免造成凌乱

Figure 18. The combination of yellow and gold reflects a sense of glory and elegance. But in actual practice, it needs to be used with a neutral colour palette.
图18. 黄色和金色的搭配富丽堂皇，但也不宜过多，在家居中还需要部分中性色调和。

residences, especially in castles built in the 18th and 19th century. White panels decorated with gold lining used with furniture of golden palette contribute to creating a glorious interior. In normal houses, wall panelling can be replaced with paint matched up with a small amount of gold lining. White and light blue is also an ideal Neo-classical palette which generates a bright and welcoming atmosphere. Usually, the colour of blue is used on walls, while white is for the ceiling, bedding and wardrobe to avoid a monotonous impression. The combination of dark red and gold is preferable for a solemn effect. (See Figure 18)

3.2 Dimensional Design in Neo-Classical Homes

The interior of a residence is composed of the horizontal and vertical dimensions whose size and shape have a direct influence on the overall effect of the space. The design of these dimensions varies according to specific position and function, but with the basic principles including fire resistance, insulation, durability, aesthetics, and continuity. Flooring design should follow a unified pattern at the entrance, in the dining room and kitchen to avoid a chaotic impression. In Neo-classical interior, wood floor and carpet provide a warm and luxurious feeling. Geometric pattern and parquet floor are the favourable option. In places like dining rooms and kitchens where there is higher hygiene requirement, ceramic tiles could be used instead of wood flooring for the surface is smoother and easier to clean. But it is essential to make sure that the tiles match the walls and the rest of the flooring in terms of colour and texture. Wood floor is always the first choice for private places of bedrooms.

Walls are the experimental field of multiple design approaches and a condensed reflection of the design concept, especially in public spaces of sitting rooms. At the beginning of the Neo-classical movement, symmetrical wood panels were a popular wall decoration. Nowadays, wallpaper is more frequently chosen and those with vine, floral or abstract patterns and thick glossy texture add to the delicate artistic features of Neo-classical style. (See Figure 19) Other than wallpaper, decorative elements are also essential for a complete Neo-classical wall decoration. As in large public spaces like sitting rooms and dining rooms, Classical paintings or tapestry of appropriate sizing could be used while in private spaces of bedrooms, wall decoration should be carried out according to the owner's preferences. Fireplaces are suggested if there is enough space. Whether in traditional or decorative style, a fireplace in the central area reminds people of

的感觉。新古典风格的家居一般会采用木制地板和地毯的搭配，既有华贵感又有温暖的感觉，地板中拼花和规则几何图形是最佳的选择，搭配的地毯要与周围的软装饰协调。在餐厅厨房这样对清洁度要求更高的地方，有时候也会应用仿古砖来代替地板，色彩质朴的仿古砖可以有效地成为木制地板的延伸，却比地板更易清洁，但需要注意的是，这种拼接的仿古砖需要在色彩和质感上与其他位置的地面以及墙面融合，共同打造出新古典的气氛。而像卧室这样的私密空间，地板是不二的选择。

Figure 19. Well proportioned geometric wall pattern presents an apparent sense of order, which enhances the diversity of Neo-classical art when arranged with arched door.
图 19. 按比例分割成几何图形的墙面有明显的秩序感，和拱形门搭配在一起更突出了新古典风格的多样性。

墙面空间给人的感觉更直观，形式也更多，是体现设计思想的重要区域，尤其是在客厅这样的公共空间，墙面的造型尤为重要。在新古典风格兴起之初，人们常用木制镶板来塑造墙面的立体对称效果，如今，壁纸是更合适的选择，藤蔓、花卉以及抽象的几何图案，加上厚重的质感和一定的光泽度，壁纸可以塑造出新古典风格所要求的精致高贵，同时也更环保（图19）。除了壁纸之外，墙面上也不能缺少必要的装饰，客厅餐厅这样面积较大且人员往来密集的地方，可以装饰古典油画、肖像画、抽象画等，或者大小合适的挂毯，而卧室这种私密性较强的地方则可以根据主人的喜好选择。当然，在空间充足的地方更不能缺少新古典风格必备的壁炉，无论是真火壁炉还是装饰性壁炉，一般都会位于空间的中心，壁炉上方可以装饰油画或镜子，壁炉框上也可以摆放工艺品和烛台。壁炉可以称得上是欧洲室内设计上的文化符号，所以在住宅中是必备的元素。

在立面中，门窗也是设计中的重点。如果需要强化新古典风格，可以使用拱形的门窗，并在门框和窗框部分做木制的雕刻。

住宅的顶面设计一般不需要过于复杂，在一些敞开式空间中可以有一些造型上的变化以区别各功能区，造型的设计也要结合灯具的选择。家居设计中普遍会选择石膏的吊顶，为了配合新古典风格，可以用石膏板把顶面做均匀的分割，或是雕刻简单的图案。另外，顶面的辅助装饰也很重要，一般在顶角线都会雕刻精致的石膏装饰性，装饰线一般是花朵图案或蔓藤图案，有时也会使用木制的装饰线（图20）。灯具的选择在顶面设计中也很重要，新古典风格对灯具的选择很宽泛，只要搭配得当，很多类型的灯具都适用。例如，具有设计感的蜡烛形吊灯、水晶灯、铁艺灯，只要根据住宅内的环境选择，就可以产生好的效果。

3.3 新古典家居中的家具陈设原则

家居空间中的家具和陈设不仅仅是家居生活中的使用工具，也是单独的艺术品。尤其是一些具有历史感的家具陈设，在一定程度上也代表着当时的生产力发展水平和艺术水平。家具在室内与人的活动关系最密切，使用频率最高，因此，实用性是选择家具的重要原则，其次，家具也是体现室内艺术效果的主要元素，所以，除了考虑符合人的使用习惯之外，也要在造型、色彩等方面符合审美要求。

新古典风格的家具秉承了古希腊和古罗马的传统，一改洛可可风格的曲线和繁复，以简洁明快的直线条为主，色彩偏重于暖色系或素淡的色彩，并且喜欢用嵌花贴皮等工艺来体现家具的华贵感。雕饰的图案主要有玫瑰花、丝带、花束等，并没有过于密集的装饰，而

Figure 20. Exquisite skirting line
图20. 精美的线脚雕刻

Europe in its heyday.

In elevation, doors and windows also count as an important compositional element. Arched doors and windows together with wood carving frames enhance the overall Neo-classical decorative style.

Ceiling design tends to be simple and elegant. Changes in design approaches are only encouraged in open spaces to distinguish different zones. The choice of lighting devices also needs to follow specific design pattern. Plaster false ceiling and exquisite plaster moulding are both quite popular in Neo-classical houses. On the contrary, a range of lamps and lanterns are available. Candlestick chandelier, crystal lamps, wrought iron lamps... are all acceptable as long as they help create a harmonious interior. (See Figure 20)

3.3 Furnishing Design in Neo-Classical Homes

The furnishings are more than tools in common households, but should be art pieces by themselves. It is especially true for historic furnishing, which indicates the productivity level and artistic development of the time when it was produced. Since they are closely related to human activities, functionality is a priority in

是追求整体的比例合理、造型精炼以及做工的考究。与之相搭配的织物也要有相应的质感，有时也会使用碎花、格子这样的图案（图21）。

现代住宅中使用的家具种类比其他空间更多，也更细致。常用的有沙发、茶几、书桌、餐桌、椅子、床、衣柜等。新古典风格对这些家具的要求并不是一成不变，无论是法国新古典的精致浪漫，还是美式的粗犷自由都可以随意地结合。而随着社会的发展以及人们对典雅生活的追求，以往只在酒店中出现的一些耐久性更高的家具开

Figure 21. Used with luxurious fabrics, Neo-classical furniture of perfect outline and balanced proportion meets both needs of function and form.
图 21. 新古典家具拥有完美的线条、合适的比例，再加上华贵的织物，兼具了功能和形式的双重要求。

Figure 22. Fireplace is the heart of a living room. It is normally decorated with art crafts and plants to add fun and sense of space.
图22. 壁炉是客厅中的中心，壁炉架上的小摆件自然不能缺少，适当的饰品再加上几种植物，空间中既有情趣又显得有条不紊。

choosing furniture. Besides, furniture needs to reflect the aesthetic and artistic personality of an interior, so shape and colour also should be taken into consideration together with the users' lifestyle and habits.

Neo-classical furniture inherits the essence of ancient Greek and Roman decoration and adopts a concise elegant concept of warm or neutral colour palette delivered via inlay/overlay techniques. Roses, ribbons, bouquets are common patterns in decorative carving, valuing balanced proportion, refined modelling and exquisite craftsmanship.

始慢慢走入家庭，并且成为一种潮流，这使得民用家具的范围也在改变，而对本来就有着开放态度的新古典风格来说，只要符合简中有繁、古中有新的精神以及开放包容又一丝不苟的态度，形式的改变并不影响它的内涵。

陈设艺术也是家居设计中不可缺少的部分，除了家具之外，织物、灯具、艺术品、绿色植物的挑选、摆放等问题也都影响着家居环境。尤其是在客厅、餐厅这样的共享空间中，需要一两处重点的陈设奠定空间的基调。在新古典风格中，壁炉上面的油画或镜面一般都是空间中的中心，油画的内容可以是风景、人物肖像等。而其他

Fabrics used on such furniture should come in matching texture. Floral and grid designs are sometimes acceptable. (See Figure 21)

There is a larger and finer variety of furniture in modern residences, most common items of which include sofa, tea table, desk, dining table, chair, bed, and wardrobe. Whether French style or the American way, the furniture pieces can all be employed in a Neo-classical home. Those with higher aging quality that used to belong in grand hotels start to appear in normal households these days, changing the trends in Neo-classical furniture and establishing a new attitude of Neo-classical home decoration.

As an essential part of a successful home design, furnishing together with the elements of furniture, fabrics, lamps, art crafts and greenery has a direct impact on the residential environment. Central furnishing pieces are especially important for public spaces like sitting room and dining room. In Neo-classical interior, paintings or mirror frames on top of fireplaces are normally the central pieces of the space. Paintings are often landscapes or portraits. Other furnishing of greater flexibility is acceptable as long as it meets the house owner's personal preference, such as sculptures, table lamps, vases and plants. (See Figure 22) The choice of fabrics requires more attention as it should be both artistic and comfortable, and thus silk, velvet and wool are preferred for their soft and luxurious texture. In private spaces, furnishing design should be focused on building privacy and personality, emphasising a cosy ambiance which is created mainly through the use of fabrics. Art crafts of personal, cultural or artistic values are all acceptable as furnishing pieces, yet the principle of focus and simplicity should be observed to create a harmonious living environment.

4. Modern Technologies in Neo-Classical Homes

Devices and material of modern technologies have become an indispensible part of every household, whichever decorative style is chosen. Apart from electric appliances, smart residential systems have been applied in grade houses or villas, including the functions of lighting control, curtain control, remote control, alarm control, and home cinema. These technologies offer more convenience to modern living and bring greater challenges to the creation of Neo-classical style. For designers, the integration of classics and technology is most important for a successful Neo-classical residence.

地方的陈设品则有很大的灵活性，有特色的雕塑，复古的台灯，精致的花瓶，以及各种绿植都很适合，只要雅致大方并符合大多数人的喜好都可以放进公共空间中（图22）。对织物的要求相对较多，因为织物除了美观之外更要有舒适性，尤其是新古典风格对织物的材料和质感有着很高的要求，一般会选择绸缎、天鹅绒、羊毛等质感华贵的面料。在私密的个人空间中，陈设艺术要围绕着私密性和主人的个性选择，强调温馨的效果，尤其是织物的选择，在很大程度上决定了空间的气氛，所以更应该选择柔软舒适且富有意境的棉、麻等材质。一些艺术品可以根据主人的爱好、经历等选择，无论是有趣味性、文化性或是有纪念意义的摆设都可以纳入其中，只是不宜品种过多，过于杂乱，尽量创造一个轻松愉悦的个人环境，同时也要考虑与新古典环境的和谐。

4.新古典家居的现代形式

无论何种设计风格，家居环境都离不开各种现代的设备和材料，除了家庭中常用的电器之外，一些高档住宅或是别墅中还会带有一些智能系统，如智能的灯光控制、窗帘控制、远程控制、报警系统、家庭影院系统等，这些设备为家居生活带来便利的同时，也对新古典风格提出了挑战。古典和现代的融合成为了设计中最重要的部分，也是设计师们最关心的部分。

在如今的家居设计中，现代化的材料也越来越常见，尤其是烤漆玻璃、塑料等类型的材料。这些材料使用方便、经济，也有很强的可塑性，经过恰当的设计，同样可以在古典的环境中找到位置（图23）。例如玻璃的应用，加上一定的镶嵌，喷砂，或加上古典的压花，可以应用在很多地方，例如在玄关位置做隔断，既可以保证空间的完整性，又能起到分割的目的。也可以用在家具中，用作家具的柜门，或是桌面，使家具带有一定的通透性，也活跃了空间的氛围。而像电视、电脑等较大又不能缺少的设施，在设计中可以尽量的弱化，加强背景的设计，并且尽量让这些设施和背景融合，或者加以遮挡，暗藏在背景中。对于灯具、开关等小型设施，则可以灵活地设计成古朴的形式，让他们成为细节中的亮点。

新古典风格需要装饰的秩序感和层次性，但也同样有着开放的精神，本书每个新古典风格的家居设计中，都包含了一定的现代设施，从中可以看出，只要有巧妙的设计，古典和现代完全可以在一个环境中共存。

Materials like paint glass and plastic have gained larger proportion in home design practice nowadays. They are economic, easy to use, and flexible. Once properly designed, these materials could as well contribute to establishing a classical environment. (See Figure 23) With inlay, sand blasting and embossing techniques, for example, glass could be applied in a Neo-classical home as glazed walls, cupboard doors or table surface. For large devices like television and computers, designers could strengthen the background in order to reduce their visual volume or keep them hidden in the background setting. Simple forms could be adopted for smaller devices like lamps and switches to make them the highlight of the ornamental details.

The sense of order and hierarchy, together with the spirit of openness, is a major feature in Neo-classical decoration. There is certain form of modern facility in every case in this book, which can be seen as a modern interpretation of Neo-classical style.

Figure 23. Glass cover on traditional fireplace provides simple protection.
图 23. 精心雕刻的壁炉用了现代的玻璃罩，简洁安全。

The art of home design exists only when functional requirements are met.

A comfortable home space can be only created on the basis that the functional requirements are realised. In a residence, the basic functional requirements include planning of spaces for resting, dining, bathing, relaxing, learning, and entertaining. The soul of home design lies in the way these spaces are arranged and created, and that's the primary issue the designer faces in a home design project.

- In a limited area, how to make a reasonable layout? Is there any standard to be reasonable?
- Nowadays when privacy gains more and more concern, what kind of a home environment can make family members feel safe and private?

家居艺术从满足功能开始

一个舒适的家居空间的打造首先要依靠功能的实现。在一个住宅空间之内，要实现家庭成员的休息、进餐、卫浴、娱乐、学习、会客等活动，这些空间功能如何布局，如何实现，是设计中的灵魂所在，也是设计师首先要解决的问题。

在一个有限的空间里，什么样的空间分布是合理的？合理的标准是什么？

在越来越重视隐私的今天，怎样的环境能让家庭成员有更强的安全感？

家庭成员之间有不同的活动，什么样的设计能让彼此不互相打扰又有人情味？

如何利用那些小面积的辅助空间，让这些地方不再平庸？

对于那些或大或小的空间怎样才能确立一个合理的形式？

如何根据空间性质确定它的开敞或封闭程度？

CHAPTER 2
FUNCTIONS OF NEO-CLASSICAL HOME DESIGN
新古典家居的功能艺术

- Family members may differ greatly in terms of daily activities, and how to make a balanced design by which different lifestyles won't be bothered while the share of love and care can be felt?
- How to make use of small auxiliary spaces and make them more than just ordinary supporting areas?
- How to find an appropriate form for large and small spaces?
- How to choose the right extent of enclosure or openness according to the function of a space?
- What is a user-oriented home environment? How to maximally satisfy users' requirement?

The satisfaction of functional requirements is not the only goal of the design, but is surely a primary objective. Home spaces usually have multiple requirements, and the satisfaction of these is the basis for the creation of a good home environment.

什么样的家居环境是以人为本的环境？怎样才能最大程度地满足人的需求？

功能的满足不是设计的唯一目的，但却是首要的目的，而家居环境的功能往往是最琐碎繁多的，满足了这些要求，也就为创造一个美好的环境打下了基础。

1. Reasonable Division of Space

Home design is about the beautification of interior spaces with physical and artistic techniques, and the division and layout of spaces is the primary issue. The division of space involves many aspects apart from architecture, such as environmental psychology and human engineering. In neo-classical style, rationality is a highlight, and thus it is particularly sensitive to the proportions of space division.

Everyone needs a territory to his own in a space, not only physically but in a psychological sense, to feel safe and private, which should be guaranteed by proper home design. Visible partitions, intangible divisions, blocking of views, and even acoustic isolation would all influence our psychological perception of an environment. Human engineering also plays an important role in home design. Human

1. 理性空间划分

家居设计需要从空间的内部入手，运用物质和艺术的手段创造出理想的环境，其中空间的划分和布局是首先要解决的问题。如何划分空间的尺度比例，涉及到许多方面，除了建筑之外，还有环境心理、人体工程等内容。尤其是新古典风格对理性的强调使其对空间划分的比例有着一定的要求。

在一个空间中，每个人不仅需要一个充足的领域，也需要一定的心理上的安全感和私密感，这些都是通过一些外部环境决定的，房间之间有形的隔断、无形的分区、视线上的阻挡，甚至声音的隔绝等都会影响人对环境的心理理解。在另一方面，空间划分也要符合人体工程学的要求和比例，空间设计中考虑大多数人的尺度和安全性，例如家具的最合适尺寸，人体活动的最佳距离，人际交往过程中的心理需求范围等，这些都直接决定了空间的大小和划分的方式。在特殊地方还要考虑针对不同的人，例如儿童房、老人房等。

dimensions and safety should be carefully considered, including the size of furniture, right distance for easy movement of human body, and the psychological requirements for space in interpersonal communication. All these are decisive factors that would influence the space scale and the way of division. In particular, some special areas for certain age groups should be designed with careful consideration, such as children's room and room for the elder.

The living room and dining room are where the neo-classical style is most obviously presented. These spaces are relatively large. In a neo-classical living room, U-shaped sofas are often found symmetrically placed around the fireplace, and ample spaces should be left for easy walking. Thus, such a living room would require a larger space than an ordinary living room. The dining room would be used for entertaining sometimes, so the dining table is often designed to sit more than six people. The bedroom should feel safe and private, and thus be relatively small. In a neo-classical bedroom, apart from the bed, usually there would be symmetrical bedside tables, a footstool at bed end, and sometimes single sofas. Enough space should be left for these pieces of furniture. In the hallway and passageways, particular attention should be paid to the wall design. If panels are used, the thickness should be proper in order to avoid overcrowding in these small areas.

The division of spaces should be reasonable in terms of not only architecture, but also the daily activities of different family members. Everyone's requirements should be met, and the spaces should be taken full advantage of. Communication between the designer and client is absolutely necessary. In the following project, the client bought a flat first, and in the process of design, bought the neighbouring one. This is the family of a young couple, and new family members are expected. So extension was required, and the two flats were combined into one, which was further divided into two main parts: private part and guest part, separated by a beautiful door. The guest part includes a bedroom, wardrobes, a bathroom and a study. The private part consists of a large living room, a dining room with kitchen, a music room which is at the same time a home theatre and can be adapted to be a nursery, a storage room and a laundry room. Together the two parts can satisfy all the requirements of the client and are adaptable to future needs.

在新古典风格的家居环境中，客厅、餐厅这样的公共区域是体现风格最明显的地方，也需要较大的环境。新古典风格的客厅常采用U形的对称沙发围绕着壁炉的形式，并且还要保留一定的过道距离，比一般现代形式的客厅需要更宽敞的空间。餐厅有时也会起到宴客的作用，所以餐桌一般都会设置六人以上的座位。卧室需要较强的私密感和安全性，所以不宜过大，新古典风格的卧室里除了床之外，还会有对称的床头柜、床尾的矮塌，有时还会有单组的沙发，所以需要给这些家具留有充足的空间。而门厅、过道等位置除了基本的要求之外，如果墙面是用镶板装饰，也要考虑镶板的厚度，以免造成过度拥挤。

一个合理的空间划分，也要根据建筑的情况，使各区域的布置符合家庭成员的生活规律，满足每个人的不同要求，同时也要保证空间的充分利用。这些需要设计师跟业主的充分沟通才能保证。在本案例中，业主先买了一间公寓，在设计过程中，又买了相邻的公寓。这是一个由一对年轻夫妇组成的家庭，但未来还会有新的成员住进来，因此他们想扩大面积。设计师把这两间合在一起的公寓重新分成了两个主要部分——个人空间和客用空间，并且用一扇漂亮的门来分割。客用空间包括一间卧室、衣柜、一间浴室和一个书房。个人空间有一个大客厅、带有厨房的餐厅和一间音乐室，这间音乐室同时也是家庭影院，而且它还可以根据需要变成一间育儿室，另外还有储藏室和洗衣房。这两个部分充分满足了业主的需求，也考虑了未来的发展。

1. Living room in Kutuzovskaya Rivera apartment
1. 库图卓弗斯卡娃里韦拉公寓起居室

Kutuzovskaya Rivera
库图卓弗斯卡娃里韦拉公寓

The new project of architect Irina Moskaleva is a presentable interior placed in one of the most elite housing estates of the capital of Russia.

From a drawing room exists an exit to the hall. In the varnished show-window by Angelo Cappellini takes place a collection of the French crystal – pride of owners of the apartment.

The owners wanted to make the interior look not "very new", but look as if having been resided by many generations. The architect decided to emphasise materials simulating an old tree on walls and scratch damages on the marble floor.

The bedroom is sustained in a warm golden-beige tone that makes atmosphere glamourous and cosy.

The bathroom looks Roman because all walls and floors are trimmed by marble entirely.

The centre of the drawing room is a library with the built-in TV and a music centre.

Walls of the hall are trimmed by panels from a nut tree. Irina took classical furniture from different collections in order to make the impression that the interior has been resided for a long time by many generations of residents. Floors are laid out by beige Crema Marfil marble and brown Dark Imperador marble with antique effect.

As it was accepted in ancient houses, the architect divided the apartment into two zones – private and guest. The owners wanted to make the interior look like solid one and at the same time cosy, as if it were created many years ago. Parquet is used on the floor with an antiquarian effect. For solemnity a stucco molding with gilding is used as well as component parts made from old bronze.

Location Moscow, Russia
Designer Baharev&Partners
Photographer Zinur Razzutdinov
Area 251.5m²

项目地址 俄罗斯，莫斯科
设计师 巴哈莱夫联合设计公司
摄影师 金努尔·拉祖迪诺夫
项目面积 251.5平方米

2. The hall with classical furniture
3. Drawing room
4. The drawing room has a music centre.
5. From a drawing room exists an exit to the hall.
6. The drawing room next to the kitchen

2. 门厅中摆设古典风格的家具
3. 客厅
4. 客厅中央配有音箱设备
5. 客厅中有通往门厅的出口
6. 客厅与厨房相连

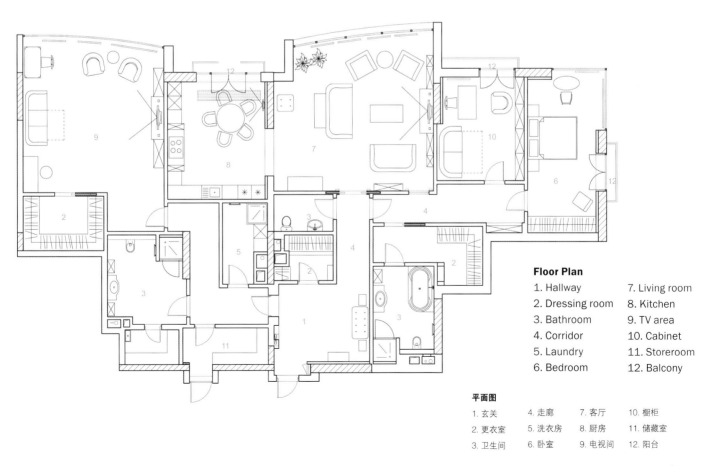

Floor Plan

1. Hallway
2. Dressing room
3. Bathroom
4. Corridor
5. Laundry
6. Bedroom
7. Living room
8. Kitchen
9. TV area
10. Cabinet
11. Storeroom
12. Balcony

平面图

1. 玄关
2. 更衣室
3. 卫生间
4. 走廊
5. 洗衣房
6. 卧室
7. 客厅
8. 厨房
9. 电视间
10. 橱柜
11. 储藏室
12. 阳台

新古典家居的功能艺术

这是建筑师伊丽娜·莫斯卡列娃的最新项目——非常漂亮的室内设计，该项目位于俄罗斯首都某社会名流住宅区中。

客厅中有通往门厅的出口，在安吉洛·卡佩里尼设计的橱窗中展示着该公寓主人珍爱的收藏品——法国水晶。

公寓主人想让室内看起来像是经历过世代的居住，而不要看起来非常新。建筑师决定着重于材料的运用，墙面上模仿了古树的图案，有的房间地板上使用了划痕大理石，还有的房间地板使用的是古香古色的镶木地板。

卧室采用了温暖的金黄色，营造一种充满魅惑而又温馨舒适的氛围。

浴室中所有的墙壁和地面都是由切割的大理石装饰的，看起来像各种罗马术语。

客厅中使用了复合材料进行装饰，中央也是一个小型图书室，装有嵌入式平板电视和音响设施。

门厅的墙面以坚果树的木板进行装饰。为了营造一种经历过世代居住的效果，伊丽娜选用了不同风格的古典家具。为了渲染一种古香古色的效果，地板采用了克丽玛米黄色的大理石和棕褐色的皇家大理石。

由于是以一种古宅的风格打造的，建筑师将公寓分成了两个区域——私人区域和客用区域。房屋主人想让室内看起来坚固而又舒适，好像是已经打造了多年的房屋。为了营造一种严肃的气氛，还使用了镀金的灰泥模塑，所配零部件均为古青铜制品。

7. The kitchen floor is laid out by marble.
8. Dining room
9. TV area
10. The bedroom is sustained in a warm golden-beige tone.
11. Master bathroom
12. Guest bathroom
13. The bathroom is decorated by marble entirely.

7. 厨房地面采用大理石
8. 餐室
9. 电视休闲区
10. 卧室采用温暖的金黄色
11. 主浴室
12. 客用浴室
13. 浴室中全部用大理石装饰

2. Safety and Privacy in Home Spaces

Psychologically men aspire for safety and privacy, so in living spaces it is not the larger the better. Instead, we need something to rely on, such as walls and columns. Privacy is particularly important in home spaces, especially in sleeping and bathing areas, where a private space with suitable isolation of light and sound is needed. In this sense, in home design you should take into account the size and shape of spaces, acoustic qualities and texture of the materials used, human perceptions and views, and requirements for different levels of privacy in different spaces. For example, a bedroom is usually no larger than 20 square metres; a broad bedroom would likely make one feel uneasy. On the contrary, a living room should be large enough to accommodate furniture and furnishings. The living room is where the design style is fully demonstrated. It usually serves

2.家居中的安全感和私密性

人的心理普遍存在着对安全感和私密性的需求，所以日常生活的空间并不是越宽敞越好，而是要有可以依靠的东西，例如柱子、墙等。同样，私密性的要求也很重要，尤其是家居生活中，人们喜欢受到尽可能少的打扰，而像睡眠、盥洗等需求更需要隐秘的环境以及光线、声音等的隔绝。从这个角度出发，在家居设计中就要充分考虑到空间的大小、形状、所用隔断材料的隔音情况、质感、人的视角、可视距离、不同功能空间中人对亲密程度的需要等问题。例如，卧室的设计一般以不超过20平方米为宜，过大就会使人产生不安的情绪。客厅则需要宽敞的空间，客厅是风格设计中的重点，需要摆放的家具、陈设品等物品很多，有些客厅还兼具待客、视听、娱乐、学习的功能，需要一定的弹性空间。而像浴室这样的空间更需要绝对的私密性和很高的隔音要求，但也要考虑人在其中活动范围和设施使用的最小余地。

1

many functions such as entertaining, accommodating audio-visual facilities, recreation and learning, and the design thus has to be flexible. The bathroom has the highest requirement for privacy and acoustic isolation, and movement of human body and use of facilities should be considered.

If the size or perception of a room is not suitable for its function, we can flexibly divide or combine spaces. Neo-classical home spaces are usually large with multiple design elements, and division or combination of spaces should be done without making the space feel empty or crowded in disorder. In particular, in neo-classical style, fabrics are often adopted, particularly heavy ones with good textures. They are helpful to the protection of privacy. You can easily find heavy window curtains or ornaments made with draperies in neo-classical home spaces. Columns are also a common element for neo-classical style. When there are enough spaces, columns can be adopted as an indispensable element of neo-classicism. They are obvious icons of neo-classical style, and can serve as a flexible technique for space division and enhancement of the sense of safety. In large-scale villas where the ceilings are particularly high, the utilisation of columns brings out psychological comfort in an otherwise too open and empty space, and provides a kind of visual shelter in the interior.

In the following project, the interior is divided into three parts. Each corridor connects to different spaces, creating a well-connected space network without interference between each other. The large dining room and the living room are connected. In the corridor leading to the dining room, delicately carved columns are placed. In this way, the wide corridor won't feel dull or empty, and views into the open dining room are properly sheltered. Master bedroom, study and guest room are placed farthest to the main corridor. The bedroom is not very big, and when the bed, sofa, desk, dressing table and necessary passageways are arranged, there's not much space left. In each room, heavy window curtains are adopted to guarantee privacy. In this home environment, division of spaces and complicated decoration are the main ways to realise safety and privacy, and are also key points of the design.

如果空间的尺寸、视角效果等因素不符合功能要求，可以人为地增加灵活的隔断或把空间重新组合。新古典风格的家居空间一般较大，元素也比较多，更需要丰富空间的层次，使空间不显得杂乱或空旷。另外，新古典风格对织物的运用很丰富，并且织物的尺寸一般较大较厚重，也有很好的质感，在一定程度上对保证私密性起到了很大作用，例如，新古典风格中常用的厚重的窗帘以及一些帷幔做成的装饰。新古典风格另一个突出的特点是柱子的使用。在有充足空间的前提下，柱子是新古典风格中不可缺少的元素，它不仅体现了新古典风格的突出特征，也能在视觉和心理上起到划分空间，增强安全感的作用。尤其是对那些面积很大、举架很高的别墅，柱子的使用能给人心理上的依托，避免了大空间带来的空旷感。也会在视觉上带来一定程度的遮挡。

在本案例中，室内的空间被严格的分成了三个部分，每一条走廊连接不同的空间，互不干扰，也能有所联系。其中大餐厅和客厅连在一起，在通往大餐厅的走廊和餐厅的门口还有精心雕琢的柱子装饰，避免了宽大走廊的空旷感，也为通透的餐厅阻挡了一部分视线。主人的卧室、书房以及客房都被安排在了离主走廊最远的地方，卧室的面积不大，在摆放了床、沙发、书桌、梳妆台和必要的过道之后，并没有太大剩余空间。每个房间都有质地厚重的窗帘遮挡，保证了房间的隐密性。在这个环境中，空间的分配和繁复的装饰是保护家居环境安全和私密的重要内容，也是设计中的重点。

1. Dining room in 320m² apartment

1. 320平方米公寓餐厅

320m² Apartment
320平方米公寓

The hostess' wish was to see the new house as "the garden of Eden" so that it would be possible to arrange gala parties and to live comfortably for a big family of six people.

The architect divided the initially open free space into three functional parts: from the threshold one mirrored corridor leads to a spacious kitchen and the owner's study; the other, located opposite, leads to the rooms, lavatories and the children's dressing rooms; and the main gallery leads to the owners' private zone and the united living room and the grand dining room. The scale of the interior design influenced not only the strict logic of the zoning but the choice of the decoration materials and the style-forming elements. The front entrance to the public zone is decorated with Corinthian columns and fountains, crystal chandeliers, and on the floor you can see a mosaic panel made of marble fragments that changes into shining black labradorite. This natural stone with its delicate texture became the main decorative device: the walls and columns of the main gallery are painted by hand imitating marble, the kitchen working surfaces are decorated with rare granite, the restrooms for guests, children and the owners are "chained" in onyx in different colours. Even the washing room is faced with marble of mustard and sand colour.

The interior of the grand living-room is the copy of Italian palazzos of the Baroque epoch. On the one side you can see the dining room unit for twelve people; on the other – traditional for classical style soft velvet sofas. The walls are decorated with silver wallpaper with golden stamp images. The combination of warm and cold colours helps avoid the effect of "burning" which is often caused by too many golden details. This method gives the richly decorated interior of the main room a feeling of bliss and sybaritic peace.

If the most part of non-residential premises are characterised by the cold shine of the natural stone, the residential ones feature a special choice of wood used for the floor decoration. The palisander parquet of the living room is edged with ornament of pear, Karelian birch and wenge; with padouk and palisander in the owners' study and with kingwood in the owners' bedroom. Material is what defines the status and the level of the customers' needs. But it is not everything: harmony, proportion and individuality are the main features of the proper interior design performance which makes even the most functional and practical presence extremely beloved and precious.

Location Moscow, Russia
Designer Anna Kulikova & Pavel Mironov
Photographer Zinur Razzutdinov
Area 320m²

项目地址 俄罗斯，莫斯科
设计师 安娜·库里克娃与帕维尔·米若诺夫
摄影师 金努尔·拉祖迪诺娃
项目面积 320平方米

2. Classic soft velvet sofas in the living room
3. Hallway decorated with Corinthian columns and fountains
4. Main gallery leading to the grand dining room

2. 起居室摆放着古典风格的天鹅绒沙发
3. 门厅用科林斯式柱和喷泉来装饰
4. 主廊道通向奢华的餐厅

Floor Plan

1. Living-dining room
2. Bedroom
3. Master bedroom
4. Master bathroom
5. Dressing room
6. Guest bathroom
7. Office
8. Kitchen-dining room
9. Entrance hall
10. Laundry
11. Bathroom
12. Fireplace room

平面图

1. 客厅/餐室
2. 卧室
3. 主卧室
4. 主卫生间
5. 更衣室
6. 客用卫生间
7. 办公室
8. 厨房/餐室
9. 入口大厅
10. 洗衣房
11. 卫生间
12. 壁炉室

5. Sitting area in the dining room
6. Grand dining room capable of holding a party.
7. Spacious kitchen
8. Owner's office

5. 餐厅中的沙发
6. 奢华的餐厅可以开派对
7. 宽敞的厨房
8. 公寓主人的办公空间

女主人希望将这座新房打造成一个"伊甸园"，这样就可以在这里举办节日派对，也可以让她们这个六口之家生活的更加舒适。

建筑师将原来开放的自由空间分成三个功能区：从入口处装有镜子的走廊通向宽敞的厨房和主人书房；另一边通往客房、盥洗室和儿童更衣室；主廊道通往主人的私人空间和与其相连的起居室，还有豪华的餐厅。室内设计的规模不但影响严密的区域逻辑规划，而且对装饰材料的选择和风格形成的元素也有影响。通往公共区域的正门入口以科林斯式柱和喷泉，还有枝形水晶灯来装饰，地面上铺的大理石板是由大理石碎片打造成的黑色拉长闪光石制成。这种具有精致纹理的天然石成为了主要的装饰性装置：主廊道中的墙壁和圆柱是模仿大理石手工喷绘而成，厨房工作台以稀有的花岗岩装饰，客人、儿童和主人共用的公共卫生间中分别镶嵌着不同颜色的缟玛瑙。甚至连洗手间中都装饰有深黄色和沙子色的大理石墙面。

9. Sleeping area featuring a special choice of wood for the floor
10. Study area in the bedroom
11. Bathroom with surfaces finished with marble.
12. Restroom for guests

9. 居住区域以精心选择的地板为特色
10. 卧室中的学习区
11. 浴室表面采用大理石
12. 客用浴室

豪华起居室的室内装饰模仿了巴洛克时期意大利豪华的宫殿设计。这间起居室可以容纳12个人，摆放着传统风格的天鹅绒沙发。墙面装饰着银灰色壁纸，并印有金色图案。冷暖色调的结合避免了由太多金色元素造成的"燃烧"效果。这种设计手法使这间装饰元素丰富的室内空间给人一种幸福、安逸与平和的感觉。

如果说大部分非居住区域是以冷色调的天然石为特色的，那么居住区域的特色就是以精选木料作为地板的装饰材料。起居室中的黑黄檀木制的镶木地板以梨树木、卡累利阿桦木和崖豆木镶边；主人书房地板以紫檀木和黑黄檀木镶边；主人卧室地板以西阿拉黄檀木镶边。装饰材料是按照客户要求选择的，体现了客户的身份地位。但是这并不能代表一切：协调、比例和个性是体现室内设计合理性的主要特征，这些特征可以将室内空间的功能性和实用性表现到极致，令人备加珍爱。

3. Private Territories for Family Members

Human beings live with sociality as well as individuality. In the indoor daily life, we try to avoid interference or invasion from the outside. When conducting different activities, we need a certain psychologically private territory. In a family, different members have different perceptions of space, and certain age groups have their own requirements. Therefore, home interior design should start from a full understanding of the requirements of all family members and create private territories according to their needs.

In a family, besides the adult couple, there might be the aged, children or servants. Sometimes extra guest rooms are needed. The master bedroom is usually placed far from the entry. The baby room, if any, should be arranged near the master bedroom. Design

3.家庭成员之间的私人领域

人们的居住需求既带有群居性，也要保持一定的独立性。人们在家庭室内的生活中，总是不希望被外界过多干扰，不希望被其他人侵犯，在家里从事不同的活动时，也有一定的心理领域。而在一个家庭中，不同的家庭成员对各个空间的感受有所不同，不同年龄段对家庭空间的需求也有所不同。所以，在对某个项目进行设计前，对每个家庭成员的需求做充分的了解，满足每个私人领域的要求，是设计师在设计过程中的重要内容。

在一个家庭中，除了年轻夫妻之外，有时还会有老人、儿童或者佣人，另外，有时也需要单独的客房。夫妻的主卧室一般会布置在离入口较远的地方，如果有单独的婴儿房，则应该靠近主卧室。设计相对简单，除了卧室之外，还会带有浴室和衣帽间，使这些功能的使用与其他成员独立出来。如果空间充足或工作的需要，还需要设

for the baby room can be simple, and apart from bedroom, there can be bathroom and cloakroom. If there is enough room or due to working requirement, separate study or work room can be planned. The master bedroom is usually shared by the couple, and thus requirements of the two should be considered. The design should conform to their common characters, not to be too feminine or too masculine, being warm and comfortable.

Design for the children's room involves many aspects, and the age and gender of the child should be taken into consideration. The room should be placed in an area with good daylight and natural ventilation. In the bedroom, apart from the sleeping area, there should be ample space for storage, play and learning. Neo-classical style generally prefers complicated furniture, but the children's room should be designed as simple as possible, and soft textiles are highly recommended. As for lighting, dazzling crystal lamps should be avoided, and ceiling lamps with soft light can be adopted.

Rooms for the aged should be designed with attention to the physical and psychological characters of the elder. The room should be placed far from frequently used spaces, in a quiet and convenient space, not to be far from the bathroom. The bedroom needn't a large area, but acoustic and ventilation qualities should be good. Soft colours and natural materials are recommended. The furniture should be safe, not being too high or too low. Psychological perception of the old people should be particularly considered. Their hobbies and experiences could be integrated into the design, if possible.

Though respective private territories are required for each family member, communication in a family is also necessary. Bedrooms shouldn't be placed far from each other. Besides, spaces for common activities are needed; for example, a large dining room, or a shared audio-visual room.

In the following project, the design brings fun to every family member. The couple share a common bedroom, but have respective bathrooms. The husband likes reading in the bathroom, so a bathroom to his own is designed with a bookshelf and armchair, while the mistress has her bathroom with a big mirror. Private spaces for their young daughter are more complicated. Apart from the bedroom, there are the wardrobe, bathroom and study. This bathroom, with decorative embossment and a copper bath tub, is the focus of the home design. Besides respective private spaces, the family members share a common space – the library. Such a design solution provides free spaces for each one, and at the same time establishes a warm atmosphere in the home environment.

置单独的书房或工作间。主卧室一般是夫妻共用，所以需要考虑两人的共同需求，设计上符合两人的共同爱好，而不宜过于女性化或男性化，以温馨、舒适为主。

儿童房需要考虑的问题比较多，针对不同年龄段和不同性别的儿童也要有不同的设计。在位置上儿童房需要阳光充足通风良好的地方，卧室除了睡眠区之外，还应该有充足的储藏区、游戏区和学习区。在家具上，新古典风格一般比较复杂，而儿童房应该尽量简单，多使用柔软的织物。灯具也不易用过于耀眼的水晶灯，尽量使用光线柔和的吊灯。

老人房的设计要更多的考虑老人的生理、心理特点，位置应该远离家庭成员活动频繁的地方，选择安静、方便的地方，离浴室也不能过远，卧室面积不需要太大，但需要良好的隔音、通风等要求。设计中尽量选用淡雅的色彩，天然的材料。家具需要更牢固、不易过高或过矮。另外，更应该注重老人的心理感受，在设计中融入老人的兴趣爱好，曾经的经历等内容。

除了各自的私人空间之外，家庭成员之间也要有交流的方式，所以，卧室之间不易过远。同时，也要有共同活动的空间，例如，一个空间充足的餐厅，共用的视听室等。

本案例中的设计为家庭成员中的每个人都带来了乐趣。夫妻俩人有共同的卧室，但在浴室的使用上，丈夫喜欢在浴室中读书，所以设计师为其提供了一个带有书架和扶手椅的浴室，而女主人则是拥有了一个带有大镜子的浴室。相比之下，小女儿的私人空间更为出色，除了卧室外，也带有衣柜、浴室和书房，而装饰着浮雕和铜制浴缸的浴室则是整个住宅中最具特色的地方。除了各自的私人空间之外，他们还有共同的领域——图书馆。这些精心的设计为每个人提供了自由的空间，也为家庭带来了温暖的气氛。

1. Bedroom in Penthouse
1. 顶层公寓卧室

Penthouse
顶层公寓

This apartment could have been anywhere. It is lucky to be in Moscow. Its design is made by a famous Russian architect and decorator Dmitry Velikovsky.

The apartment consists of several private "areas" – an old and very comfortable tradition. The boundary premises between the private areas of the family members – library hall. From here you can go to the main bedroom with adjacent wardrobe and bathroom. The English master's bathroom is designed with the English humour: if a man likes to read in the bathroom he would need a library and an armchair there.

A mirror bathroom of the mistress is, vise-versa, an example of French sophistication, as well as her boudoir separating bathroom from the bedroom. None of the objects loses its functions in favour of the style, symmetry or status. Maybe that's why the house does not make an impression of "executive apartments".

Art collection includes only pieces that really please. The best works are in the younger daughter's chambers. Her area is not inferior to the master's. Bedroom, study, wardrobe and bathroom are spacious as well. It seems that the most beautiful bathroom is decorated on the same principle – only the best: copper bath, bas-reliefs and elegant wall colours.

Behind the daughter's area there is a guest part of the house, upholstered in silk with straw patterns in combination with turquoise colour in decoration elements.

It is necessary to mention the living room with a big table, and collection of gravures in dark grey, almost black antechamber, the master's study, and the kitchen in ivory. "Dutch" still nature in the kitchen is not of the 17th century. These are works by Russian modern painters Yuri Bakrushev and Sergei Gerasimov working under pen name of Yuri German.

The impression that makes the apartment is that it is a sweet home saving memory of many generations.

Location Moscow, Russia
Designer Dmitry Velikovsky
Photographer Tim Beddow
Area 429m²

项目地址 俄罗斯，莫斯科
设计师 德米特里·维利科夫斯基
摄影师 蒂姆·贝多
项目面积 429平方米

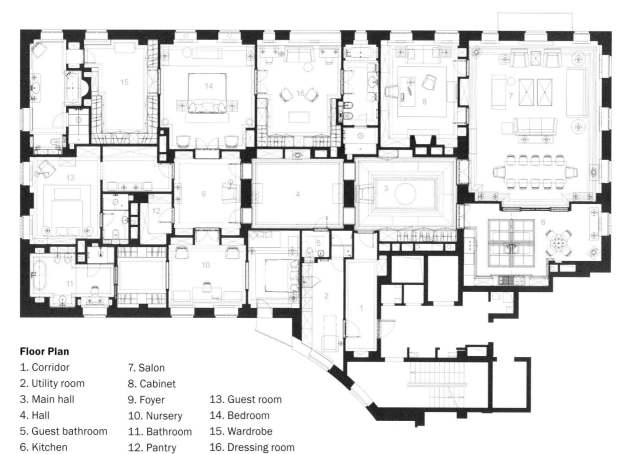

Floor Plan

1. Corridor
2. Utility room
3. Main hall
4. Hall
5. Guest bathroom
6. Kitchen
7. Salon
8. Cabinet
9. Foyer
10. Nursery
11. Bathroom
12. Pantry
13. Guest room
14. Bedroom
15. Wardrobe
16. Dressing room

平面图

1. 走廊
2. 设备间
3. 主厅
4. 大厅
5. 客用卫生间
6. 厨房
7. 待客沙龙
8. 艺术品陈列柜
9. 门厅
10. 儿童室
11. 卫生间
12. 橱柜
13. 客室
14. 卧室
15. 衣帽间
16. 更衣室

这间公寓可能在任何地方。幸运的是它在莫斯科,其设计由俄罗斯知名建筑师、装饰大师德米特里·维利科夫斯基操刀。

这间公寓的室内环境由若干私密空间构成,这是俄罗斯的一项古老传统,这样的房子居住起来非常舒适。图书室是家庭成员的私密空间之间的分界线。从图书室可以进入主卧,主卧带衣帽间和浴室。房主是英国人,所以主卧浴室的设计体现出特有的英式幽默:如果你喜欢在浴室里读书的话,摆上一张扶手椅,浴室就变成图书室了!

女主人的浴室则与男主人的相反,安装了镜面,体现出法式的精致情调,卧室和浴室之间是化妆室。每一样东西都有其用途,同时又能形成对称的格局或者营造风格和氛围。也许正因如此,这间屋子才不会给人留下"行政酒店公寓"的印象。

室内的艺术装饰品用的都是真正令人赏心悦目的东西。最好的饰品都在小女儿的卧房里。她的空间不亚于男主人的。卧室、书房、衣帽间和浴室也都非常宽敞。似乎这间最美丽的浴室也采用了相同的装饰原则——只用最好的,包括铜质浴盆和浅浮雕艺术品,墙壁采用淡雅的色彩。

小女儿的空间后面是会客区,采用凌乱图案的丝织品来布置,搭配蓝绿色的装饰元素。

值得一提的还有起居室,布置了一张大桌子,还有一套深灰色的凹版印刷画。前厅的空间几乎是黑色的,男主人的书房和厨房则采用象牙白。厨房里的画作不是17世纪的,而是俄罗斯现代画家尤里·巴克鲁谢夫和谢尔盖·格拉希莫夫的作品,二人的笔名是尤里·日耳曼。

这间公寓给人的整体印象是:这是一个甜蜜温馨的家,保有几代人的温情记忆。

2. Cooper bath and bas-reliefs in younger daughter's bathroom
3. The antechamber is almost black.
4. The mistress bathroom is an example of French sophistication.
5. Younger daughter's bedroom
6. The main bathroom also can be used as library.
7. The main bedroom has sophisticated silk.

2. 小女儿浴室中的铜质浴盆和浮雕艺术品
3. 前厅的色调近乎黑色
4. 女主人的浴室凸显法式精致情调
5. 小女儿卧室
6. 主浴室可以用作图书室
7. 主卧采用精致的丝织品装饰

4. Interior Passageways

In home spaces different rooms, stairs and hallways are connected with passageways, which also serve as important spaces for urgent evacuation. In a residence with large spaces and more rooms, the functions of passageways would be more evident. However, design for the passageway is often ignored; it usually exists only as an auxiliary space without much decoration. While in neo-classical home design, in order to achieve consistence and unity of the interior style, passageways are often carefully treated, being more than just spaces for walking, and can be designed with various forms.

Passageways in the home environment are used for walking for a few people, but enough spaces should be provided because two would be using a passageway at the same time and furniture might need

4.室内过道的处理

住宅中的过道用来连接各个房间、楼梯、门厅等部分，同时是承担疏散功能的通道。在面积较大、房间较多的住宅中，过道的作用就更为明显。在许多风格中，过道的设计常常被忽略，没有过多的装饰，只是作为辅助功能而存在。而在新古典风格中，为了保证风格的一致和统一，过道不仅仅是供人行走的通道，同样可以有多彩的形式。

家居中的过道通行人数少，但是也要考虑两人相对以及搬运家具所需要的最小尺寸，并且新古典风格在过道墙面一般会有装饰，所以过道的宽度不宜过窄，一般不宜小于1.2米。家居设计也不同于酒店旅馆，家居中的过道不应过长，或者一条过道连接很多房间，而应当有所变化。一般情况下，过道应该尽量避免有坡度，尤其是有老人或儿童居住的空间，如果必须有坡度，也要控制在一定范围，并

to be moved sometimes. In addition, in neo-classical home spaces, passageways are often decorated with ornaments, so enough width should be provided (generally no less than 1.2 metres). Different from hotels, passageways in homes should not be too long, and you should not connect many rooms with just one passageway. Instead, design for home passageways should be varied. In general, passageways shouldn't be placed on a slope, especially in living spaces for the elder and children. If a sloping passageway is unavoidable, the gradient should be well controlled, and skid resistance measures should be taken.

In neo-classical homes, design for passageways should be diversified. Dull and depressing passageways are to be avoided. Before design starts, you should leave enough space for passageways. There are many choices for decorations and treatments. Panels or wallpaper are often adopted as the basic treatment. If the panels span a large area, you can cut them with reasonable proportions and decorate them with simple patterns. If wallpaper is used, you'd better choose the one with classical patterns and luxurious textures. When the passageway is very long, besides these basic treatments, you can add extra decorations to avoid the feeling of oppression. If the passageway is too narrow, you can use mirrors on the wall to visually enlarge the space, especially mirrors with golden frames that go well with the neo-classical style. Attention should be paid to the size of the mirror; a mirror that spans too large an area is not recommended, and the frame should be harmonious with the surrounding environment. For instance, you can choose the same patterns as in other parts of the interior. Wall lamps are common decorative elements in a passageway. For the neo-classical style, it's better to choose wall lamps with exquisite details and place them symmetrically. A passageway is a three-dimensional space and apart from the walls, the ceiling and floor are also important surfaces. Usually the two surfaces would be designed in cooperation with each other. Plaster carvings are often adopted on the ceiling, and carved mouldings can be added where the ceiling and the wall meets, while marble or parquets are often used on the floor.

In the following project, the design for the passageway is a highlight. The curving passageway creates a dynamic space, with dark red tone and symmetrical wall lamps completing a typical neo-classical style interior. Such a design, simple yet delicate, made the passageway a warm and comfortable space. In the alcove, a sofa is placed, with the same material texture as that of the wallpaper, further enhancing the soft, cosy atmosphere. This solution is particularly suitable for large passageways; though difficult in implementation, the final result is surprisingly good.

采取一定的防滑措施。

对新古典风格来说，过道的设计应该更丰富多彩，避免沉闷，并且在设计前，首先为过道预留充足的空间。在装饰上，可以有很多选择。一般会在墙面上做镶板或壁纸的基础处理，如果是大面积的镶板，可以按比例分割，并雕饰简单的花纹，而壁纸则应选择花纹古典、质感华贵的风格。除了基本的装饰之外，如果过道较长，也需要增加额外的装饰来减少沉闷感。如果过道空间不充足，可以采用装饰镜面的方式，在新古典风格中，常用镶嵌金色框的镜子装饰墙面，用在过道中，也可以增加狭窄过道的空间感。但镜面的装饰不宜过多过大，并且镜框的装饰应该与环境和谐，例如，可以使用与其他地方呼应的花纹细节。另外，壁灯也是过道中的常见装饰，但对新古典风格来说，壁灯需要精致的造型和对称的设置。除了墙面之外，上下两个界面的设计同样重要。天棚和地面的设计通常会在造型和花纹上有所呼应，天棚一般会用石膏雕刻花纹，并在与墙面的连接处雕刻线脚，而地面则会采用大理石或实木的材质。

在本案例中，过道的设计极富特色。尤其是一段曲面的过道，一改平面造型的呆板，使过道的空间也变得灵动起来。再加上新古典风格中常用的暗红色调和对称的壁灯，简单的装饰让过道也成了一处风景。在曲线的凹处，依照墙面的弧度摆放了沙发供人休息，沙发的质感和墙面的壁纸浑然一体，又为这个空间增加了柔软温暖的感觉。这种设计适合较大空间的过道处理，在实施中也有一定的难度，但也起到了意想不到的效果。

1. Corridor in Nikolsky Deadlock house
2. Curve corridor

1. 尼克尔斯基公寓走廊
2. 走廊的曲线造型非常别致

Nikolsky Deadlock
尼克尔斯基公寓

Designers of this project face a difficult problem with the layout of living space in the house, which is built with a semi-circle form. As a result all the rooms are designed with a trapezoid form. The designers manage to take advantage of the existing circumstance: the non-conventional configuration of rooms allows the designers to put various pieces of furniture together, and make sliding partitions. The clients – a married couple – are not that inclined to make the interiors in classical style at first, but the designers convince them to try some new-style classics. Ivory, light grey and pale yellow tones are chosen for the interior to compensate the lack of sun light partially blocked by trees. The kitchen, dining room, drawing room and hall are filled with light which is reflected from a set of mirrors on the wall. Thanks to mirrors, impressive plasters and an inlaid floor, the hall looks solemnly extraordinary with the same tone.

The special attention is paid to the floor furnishing, and natural stone and wood are used. The ornament on the stone is based on its own different texture and meets the owners' needs of Florentine mosaic work. Thus the seam between fragments of different patterns is almost indiscernible, and the mosaic looks like a carpet. The lunch zone in the kitchen-dining room is created in the form of a circle: plastered frieze on the ceiling is accordant with the ornamental mosaic on the floor.

The incrustation is quite classical: geometrical mosaic for the working area, garland for the hall and edges, three-dimensional drafts for the bedroom, butterfly for the children's room. It is interesting to note that the inlaid medallions on the parquet and decorative plafonds in the stucco molding frame on the ceiling are the important elements used by architects to make the polygonal room more composite. The master bedroom is in empire style: a lamp with porcelain base and doors in the Greek style. The master bathroom is decorated with creamy-tone walls; the ceiling is completely trimmed with wooden panels.

Location Moscow, Russia
Designer Dom-A
Photographer Dmitry Livshits
Area 270m²

项目地址 俄罗斯，莫斯科
设计师 Dom-A设计工作室
摄影师 德米特里·列夫谢茨
项目面积 270平方米

Lobby Level Plan
1. Hallway
2. Dressing room
3. Laundry
4. Cabinet
5. Corridor
6. Kitchen
7. Dining room
8. Living room
9. Hall
10. Children and games
11. Bedroom
12. Bathroom in the bedroom

大堂平面图
1. 玄关
2. 更衣室
3. 洗衣房
4. 橱柜
5. 走廊
6. 厨房
7. 餐室
8. 客厅
9. 大厅
10. 儿童游乐室
11. 卧室
12. 卧室里的卫生间

3. The drawing room is filled with light reflected from the mirrors on the wall.
4. The lunch zone in the kitchen-dining room
5. Cabinet in the office
6. Natural stone is used in the corridor.

3. 客厅洒满镜面墙上反射的光线
4. 餐厅和厨房中的用餐区
5. 办公室里设置了橱柜
6. 走廊地面采用天然石材

本案的设计师面临的难题是：这栋建筑的空间形态很特殊，是半圆形，里面的所有生活空间都是扇形的。因此，所有房间都呈现出不规则四边形的格局。设计师巧妙地利用了既定条件，即空间的非常规形状，将各种家具灵活地组合，并采用了推拉隔断。委托客户——一对已婚夫妇——最开始并没有打算将室内装修成古典风格，但是设计师说服他们去尝试一下新古典。象牙白、浅灰色和淡黄色是室内空间选用的主色调，目的是弥补室内自然光线的不足，因为室外的树木遮挡了一定的光线。厨房、饭厅、客厅和门厅的照明主要来自镜面反射的光线，设计师在墙面上设置了一系列的镜子。镜面、石膏工艺和镶饰地板等元素的使用，让门厅在延续了室内统一色调的前提下，看起来异常恢弘大气。

设计师尤其重视地面的装修，采用了天然石材和木材。石材上的装饰花纹都是石头的天然纹理，迎合了房主对佛罗伦萨绘画风格的倾慕。不同花纹之间的过渡非常自然，几乎看不出来，拼接的效果看起来就像地毯一样。午餐用餐区是"厨房+饭厅"空间的一部分，打造成圆形的空间，天花上的石膏工艺与地面上的拼接装饰花纹相得益彰。

室内的装饰元素都体现出古典风格：办公空间采用几何构图的瓷砖；门厅采用了花环；卧室采用了三维立体画；儿童房采用了蝴蝶图案。值得一提的是，镶木地板上的嵌入式图案以及灰泥塑形的装饰天花板，都是设计师所用的重要元素，让多边形的空间看上去更加丰富、多样化。主卧呈现出皇室的风格：灯具采用瓷器材料，门则是希腊风格的。主浴室采用乳白色的墙面来装饰，天花全部采用木板镶嵌。

070　FUNCTIONS OF NEO-CLASSICAL HOME DESIGN

7. Bedroom of the owner in empire style
8. Bedroom of the hostess
9. Master bathroom
10. Children's room
11. Butterfly for children's room

7. 公寓男主人的卧室凸显皇室风格
8. 公寓女主人卧室
9. 主浴室
10. 儿童房
11. 儿童房地板上有蝴蝶图案

新古典家居的功能艺术

5. Ample Space and Right Zoning

When the main programmes are defined, zoning should be done with consideration of the existing architectural conditions and artistic techniques to enrich the home interior, creating reasonable flow of space and a clear identity and making maximum use of the space. The process of zoning is to organise and improve functionality of spaces, or in other words, to create new spaces.

A residence may be large or small in area, and the layout may be reasonable or not very satisfying. A good design is to re-create spaces with reasonable zoning to complement the existing disadvantages. Neo-classical home design is suitable to be carried out in big houses, but not all homes can meet the requirement on space area. A right zoning can be helpful; you can define, enclose, partition or combine

5.充足的空间和正确的形式

在家居空间的分区明确后，利用空间形态、空间构图的组织和安排，加上艺术的手法的渲染，使空间层次更丰富，流线更合理，空间的形象更明确，空间利用率更高，这一过程是对空间功能的完善，组织，也是创造新的空间形态的过程。

每个家居空间有大有小，布局有合理也有不尽如人意，重新创造的过程，需要合理的形式弥补空间的缺点。新古典风格的家居需要宽敞的空间，但不是所有空间都能满足要求，通过空间的限定、围合、分割、重组才能创造出合理的形式。在一般的家居环境中，入口处的空间一般会分割出单独的衣帽间，并紧邻着客用卫生间，方便客人使用。主通道直接连接客厅，这种流线的设计最大程度地方便了来访的客人。客厅需要最大的面积，尤其是新古典风格。对一些大型别墅类住宅来说，一层有时会有大面积挑高的空间，适合用

spaces. In a residence, a cloakroom is often created at the entrance area, close to guest bathroom for the convenience of guests. The main passageway connects to the living room, making a flow of space for easy navigating for guests. The living room should be large, especially in a neo-classical style home. In big villas, there is often a high-ceiling living room on the ground floor, but in some apartments, the ground floor area is limited, and then you could place the living room on other floors or re-organise the spaces. If the living room has a small area, decoration should be relatively simple. In a neo-classical style home, the living room is often placed near the dining room and kitchen, especially the large dining room for entertaining guests. In this way, after dining, the guest can conveniently have a rest in the living room. The master bedroom is usually placed in the deepest end. A large bedroom can be divided into two parts: sleeping area and leisure space. The way of division can be flexible; you may make use of the placement of furniture, or simply use columns. In the leisure space, symmetrically placed sofas and a pretty tea table are the most common choices. If the bedroom is small, the use of furniture should be limited. A simple four-poster bed with symmetrical bedside tables can establish a satisfying neo-classical style as well.

The following project is part of a townhouse, with three floors, each having a limited area, long and narrow. The solution is that only a simple entrance lobby and wardrobes are placed on the ground floor; above on the first floor, a large area is left as the living room, and a small space is kept as kitchen; the top floor is placed with master bedroom and workspace. Although the area is limited on each floor, the programmes are relatively simple, so each zone gets ample space. The client didn't want his house to look crowded, so each room was designed with as less furniture and decoration as possible, and the interior looks bright and open. Especially in the bedroom, a leisure space is defined by a platform and two classical columns, which makes the room distinctive. A subtle division is realised with typical neo-classical elements. Full-height columns are seldom used in bedrooms, but here they are well suited with the environment. Each space has its own characteristics on the basis of right zoning.

来做宏伟的客厅，但对一些一层面积不大，或一些公寓来说，客厅的面积就需要移到其他楼层，或重新组合空间。如果客厅面积不充足，那么在装饰上就应该尽量减少，利用简洁的形式。在新古典风格中，客厅常常会靠近餐厅厨房，尤其是用来待客的大餐厅。在宴客之后，客人可以方便的进入客厅休息。而主人的卧室往往处在最里面，面积大的卧室可以分割出睡眠区和休闲区，分割的方式有很多，用家具的摆放来形成自然的分区，也可以使用简单的柱子。休闲区中，对称的沙发和一个小茶几是最好的选择。而如果卧室面积比较小，就可以尽量减少家具的使用，简单的四柱床，对称的床头柜同样可以创造出新古典风格的效果。

本案例是一栋联排别墅的一部分，共有三层，每层的面积并不大，并且呈狭长的形状。所以在一层只有简单的入口大厅和衣橱。而在二层留出了较大面积作为客厅，里面稍小的面积作为厨房，三层是主人的卧室和工作室。虽然别墅每层的面积不大，但空间类别并不多，所以每个空间都有充足的位置，但由于业主并不喜欢拥挤，所以每个房间都布置了尽量少的家具和装饰，使环境看起来宽敞明亮。尤其是在卧室中，由一个地台和两根古典柱式分割出来的休闲区使卧室变得与众不同，体现了新古典风格最重要的元素，也用最简洁的方式做了分区处理。在卧室中使用全柱式的形式很少见，但在这里，柱式的作用被发挥得淋漓尽致。每个空间都有自己的个性，只要找到符合它的形式，就可以有不一样的收获。

1. Living room in the House on the Jasmine Street
1. 茉莉大街别墅起居室

House on the Jasmine Street
茉莉大街别墅

The house is part of a townhouse, with three storeys and a loft. The ground floor is a garage, hallway and a large walk-in closet. On the first floor are kitchen, living room and guest toilet. This floor is designed for entertaining, meetings, family reunions and other events. The second floor is private zone. It contains the master bedroom, study and bathroom for the owners. The attic is reserved for the owners' hobby – there is located a music studio.

The clients are very interesting and cheerful people. They do not like interiors saturated with furniture, but attach value to low-profile spaces. The designers were asked to create a quiet classical interior, but with modern elements. Therefore, in the interior there are no sharp accents or flashy parts to provoke topics. The appearance of the home does not cause irritation. The house is good that all of its rooms are large enough and have the correct form. Sophisticated decorative elements are not guilty of excesses and pretentiousness.

The colour palette of the guest area is dominated by maroon and beige tones. To finish the living room and kitchen walls using wallpaper in different colours and ornaments, the floor is paved with granite, ceramic with a classic pattern that betrays the personality of the room. Centre of the composition is a living room fireplace. It is made of beige marble with red veins. On both sides of the fireplace are portraits of the owners of the house. Also located in the living room are a group of sofas and chairs, which fit well into the interior. The clients wanted to combine classic style with elements of modern interior. So in the living room and kitchen there are curtain lights as an additional lighting. Stucco emphasised restrained and confined ceiling cornices and pilasters. Moderation in the decoration of the walls and ceiling helps focus attention on the faces of the protagonists of the interior, fireplace and luxurious pieces of furniture.

The second floor of the house is completely given over to the private zone. The master bedroom is divided into two zones by two classical columns and a podium: a zone of sleep and rest area. In the area of sleep is a large luxurious bed. It takes central stage in this room. Mirrors are located near the bed, so the volume of the room is multiplied. In the recreation area is a table with two chairs, for a quiet and relaxing evening with wine and fruit.

The clients wanted the interior to be decorated in the traditional classical style. Like classical music, classicism has a lot of advantages. Such a house will be perceived as a family home, a family legend, a mansion. Respectable solid interior symbolises the strength of family ties and cultural traditions. In the end, it never goes out of fashion, the more that is created with durable natural materials.

Location Magnitogorsk, Russia
Designer AS 20/10
Photographer Gorbatov Ilya
Area 200m²

项目地址 俄罗斯，马格尼托哥尔斯克
设计师 AS 20/10设计工作室
摄影师 戈尔巴托夫·伊利亚
项目面积 200平方米

2. Cabinet in the kitchen
3. The fireplace in the living room is made of beige marble with red veins.
4. The sofas fit well into the interior.
5. Kitchen on the first floor
6. Dining area using wall paper with classic patterns

2. 厨房里的精致橱柜
3. 起居室里的壁炉采用带红色纹理的浅褐色大理石制成
4. 沙发非常适合室内的风格
5. 二楼厨房
6. 餐室墙面采用了带古典纹饰的墙纸

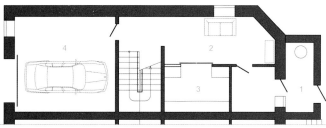

Ground Floor Plan 一楼平面图

1. Entry   1. 入口
2. Hall   2. 大厅
3. Closet   3. 壁橱
4. Garage   4. 车库

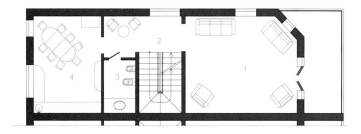

First Floor Plan 二楼平面图

1. Living room   1. 客厅
2. Hall   2. 大厅
3. W.C.   3. 卫生间
4. Kitchen   4. 厨房

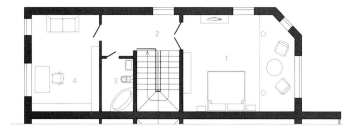

Second Floor Plan 三楼平面图

1. Bedroom   1. 卧室
2. Hall   2. 大厅
3. Bathroom   3. 浴室
4. Home office   4. 家庭办公室

7. The master bedroom is divided into two zones by two classical columns and a podium.
8. Large luxurious bed in the sleep area

7. 两根古典风格的立柱和一个平台将主卧一分为二
8. 睡眠区超大尺寸、极尽奢华的床

这栋别墅属于一系列联排别墅的一部分，共三层，带阁楼。一楼是车库、门厅和大型步入式衣帽间。二楼有厨房、起居室和客用卫生间。这一层主要用作休闲娱乐、家庭聚会以及其他各种活动。三楼是较私密的空间，有主卧、书房和主浴室。阁楼是根据主人的爱好专门打造的——一间录音室。

委托客户是非常有趣、开朗的人。他们偏爱低调的空间，不想室内家具过于繁冗。他们要求设计师打造古典风格的室内空间，但要采用现代元素。因此，室内没有特别突出的亮点，也没有炫目的俗丽，一切低调而平淡，不会引起兴奋的感觉。所有房间都很宽敞，格局合理。复杂的装饰性元素没有过度使用。

客用空间的色调以栗色和浅褐色为主。起居室和厨房的墙面采用了不同色彩和图案的墙纸，地面采用花岗岩铺装，形成古典主义的图案，凸显了空间的古典风格。起居室里的中心元素是壁炉，采用带红色纹理的浅褐色大理石制成。壁炉两边的墙上挂着房主的肖像画。起居室里还有一套沙发和座椅，风格上与室内空间融为一体。

委托客户想要将古典风格和现代室内元素相结合，所以起居室和厨房采用了别致的灯具。设计师用灰泥突出了顶棚线脚和壁柱。墙面和天花适度的装饰，让人更容易将注意力放在室内空间的主角上——壁炉和奢华的家具。

三楼主要是房主的私密空间。两根古典风格的立柱和一个平台将主卧一分为二：一部分是睡眠区，另一部分是休闲区。睡眠区有一张超大尺寸、极其奢华的床，占据了这个房间的中央。床的周围设置了许多镜面，从视觉效果上扩大了空间的体量。休闲区有一张桌子、两把椅子，傍晚时分可以在这里喝点酒、吃点水果，非常惬意。

委托客户希望室内空间采用传统的古典风格来装饰。就像古典音乐一样，古典主义风格有许多特点。这样风格的房屋更容易被视作代代传承的家宅。坚实的室内空间象征了家庭凝聚力和文化传统。最后，古典主义永不过时，越是采用经久耐用的天然材料，就越是历久弥新。

6. Open and Enclosed Areas

Open areas are extravert spaces without being strictly defined and privacy is less protected. They are flowing and penetrating, aiming at communicating with other spaces, enlarging views and enhancing flexibility. On the contrary, enclosed areas are introvert spaces, defined by enclosing entities. They are static spaces isolated from outside views and sounds, being safe and private, feeling as a territory to one's own. Nonetheless, the so-called open and enclosed areas are not unalterable, and different degrees of enclosure can be flexibly achieved, creating semi-open or semi-enclosed spaces.

The use of open or enclosed spaces depends on the function of the space, psychological needs of the user, and the relationship with the surrounding environment. In an entire home environment, the

6.开敞空间与封闭空间

开敞空间是外向型的空间，没有固定的限定性，私密性也比较小，是流动的，渗透的，它的目的是与其他空间环境交流，扩大视野，因此灵活性也很大。而封闭空间是内向型的，用限定性高的围护实体包围起来，它是静止的，可以隔绝外来的视线、声音等干扰，能给人以安全感、私密感和领域感。但所谓的开敞空间和封闭空间也并非绝对，界面的围合程度、大小等都决定了开敞的程度，有时也会出现介于两者之间的半开敞或半封闭空间。

开敞空间或封闭空间的使用取决于空间的使用性质、人在其中的心理需要以及和周围环境的关系。在一个完整的家居环境中，两种类型的空间都可以起到很大作用。一般来说，卧室、卫浴、衣帽间等私密性较强的地方需要相对封闭的空间，但根据所处的位置也可以做适当的调整，例如在卧室里面的浴室或衣帽间。而开敞空间的把

two types of spaces can be both helpful. Generally speaking, private spaces such as the bedroom, bathroom and cloakroom should be enclosed, but appropriate adjustment is possible according to specific conditions; for example, the bathroom or cloakroom in the bedroom. Design of open spaces is relatively difficult. In home interiors, it's hard to find a space absolutely open, but spaces with some degrees of openness are often needed. From the entrance hall to the living room, dining room and kitchen, all spaces could be designed with a certain degree of openness. In a neo-classical style home, the entrance hall is often large, and passageways leading to other rooms need to feel open, but commanding views of everything should be avoided. Therefore, semi-open space is the best solution. You may use soft partitions or a foyer. In a typical neo-classical home design, the dining room and kitchen are often directly connected without any partition in between for easy communication. Sometimes the dining room for entertaining guests is designed semi-open. Apart from enclosing walls, a large arch can be added, connecting it to the living room or passageway. It is common in neo-classical style home design to arrange the dining room and kitchen in one place, usually feeling more open, well-connected with other interior spaces, and sometimes even connected with outdoor landscapes, making the interior feel vibrant. Moreover, open spaces are adaptable with flexible arrangements of furniture and furnishings, thus being multi-functional in use.

In the following project, the bedroom, study and bathroom are relatively enclosed. The semicircular entrance hall opens to the dining room and living room, which are connected, forming a large open space together with the entrance hall. The kitchen is partitioned from the dining room. Besides that, two patterned mat glass doors featuring primitive simplicity are added, in coordination with the surrounding environment. In this way, the kitchen and dining room are separated yet connected in a subtle way.

In general, home interiors are limited in area, and thus the degree of openness is restricted. An appropriate play of openness and enclosure would endow the space with a sense of beauty.

握相对更困难。在家居环境中，很难找到一个绝对开敞的空间，但在很多环境中却都需要一定的开敞程度。从入口的门厅、到客厅、餐厅、厨房都可以设计一定的开敞范围。在新古典风格中，门厅一般会比较宽敞，通往其他房间的通道需要一定的通透性，又不能对室内情况一览无余，所以，更适合采用半开敞的方式解决这一问题，或采用软隔断、玄关等。新古典风格的餐厅和厨房常常会连在一起，中间没有隔断，这样能使两个空间的交流更容易。而用于宴客的餐厅有时也会有一定的开敞程度，除了周围墙壁的围合之外，增加一个大的拱门与客厅或走廊相连。餐厅和客厅在一个空间中也是新古典风格中常见的形式，而这样的空间开敞程度往往会更大，既能与室内其他空间有所联系，有时也会与室外景观相连，给人开朗、活跃的感觉。另外，开敞的空间有很大的灵活性，可以改变室内的布置方式，所以在功能上也有很大的弹性。

在本案例中，卧室、书房、浴室都相对封闭，半圆形门厅的开敞处对着餐厅和客厅，餐厅和客厅相连，不仅与门厅形成了一个开敞的空间，也与厨房相连，厨房和餐厅间有隔断，但却增加了两扇有一定通透性的印花磨砂玻璃门，这两扇门造型古朴，与周围环境相得益彰，隔绝了厨房和餐厅的同时也让两个不同空间可以随时保持通透，有所联系。

家居环境的空间有限，开敞的范围也有限，只有适度的封闭和开敞才能创造出无尽的美感。

1. Cabinet in Pokrovsk apartment

1. 波克罗夫斯克公寓特色橱柜

Pokrovsk
波克罗夫斯克公寓

The style of Marina Putilovskaya is recognisable from the doorway. Marble panels, a soft carpet, stained glass, decorative moldings, massive dark ceiling, and elegant furniture make the sparkling interior really stylish and harmonic. Every interior element is independent, but when they all come together, they create a unified sound.

Owners of the apartment are experienced in interior details and fashion trends. They are real professionals who have articulated their wishes. The main thing in this case was that the interior reflects the nature of the host and that it is not shining, "status" attributes, and perfection inherent to true works of art.

The starting point became the golden thread dining chairs Cappelletti, and because of them turned out a quite unconventional interior.

In this apartment there are many bright and very unique accents. For example, the flooring in the apartment is made of a very rare and expensive wood, which glows from the inside as amber.

Unusual is also a kitchen. When the designers were choosing a kitchen the owner asked to make it with a very rare material, made in the style of "bull". The basis here is the tortoise shell, decorated with brass inlays. Design sketches were created by Design Bureau of Marina Putilovskaya. Equally unusual is the mosaic ceiling of the room, made of fine wood in the art "intarsia".

Another masterpiece is the master bathroom. The small space required to increase the visual effect, and the solution was the marble floor of the huge roses. The combination of dark stems with watercolour petals and glittering mirror background creates an unexpected and equally brilliant effect, as in other areas of the apartment.

Each room in the apartment is built on its own circuit, carrying its own character, but at the same time maintaining a mood of the entire apartment. Creating a one-piece interior with bright accents was the hardest, but the designers made it through the experience of their offices and their partners.

Location Moscow, Russia
Designer Design Bureau of Marina Putilovskaya / Marina Putilovskaya
Photographer Zinur Razutdinov
Area 357m²

项目地址 俄罗斯，莫斯科
设计师 玛丽娜·普蒂洛夫斯卡亚设计工作室（玛丽娜·普蒂洛夫斯卡亚）
摄影师 基努尔·拉祖蒂诺夫
项目面积 357平方米

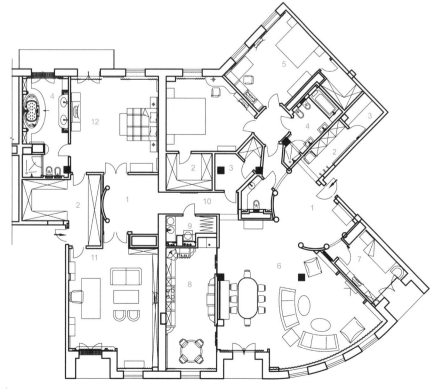

2. The door in the living room has decorated with glass
3. Living room with elegant furniture
4. The hall features marble panels.

2. 客厅的门采用彩色玻璃
3. 客厅家具精致典雅
4. 大厅采用大理石板材

Floor plan

1. Hall	7. Guest room
2. Wardrobe	8. Kitchen
3. Pantry	9. Laundry room
4. WC	10. Corridor
5. Children room	11. Cabinet
6. Foyer	12. Bedroom

平面图

1. 门厅	7. 客卧
2. 衣帽间	8. 厨房
3. 餐具室	9. 洗衣间
4. 卫生间	10. 走廊
5. 儿童房	11. 陈列柜
6. 大厅	12. 卧室

FUNCTIONS OF NEO-CLASSICAL HOME DESIGN

马利娜·普蒂洛夫斯卡亚的设计风格从门口就能一眼认出。大理石板材、柔软的地毯、彩色玻璃、装饰性线脚、大体量的深色天花以及精致典雅的家具，共同铸就了闪亮的室内空间，其风格既时尚前卫，又和谐统一。每件室内设计元素都是独立的，但是搭配在一起，却又显得那么和谐。

这间公寓的主人对室内装饰的细节和流行趋势非常在行。他们的水平已经很专业了，所以能够向设计师精确地传达出他们想要的效果。本案的设计核心是要让室内空间体现出主人的品位，但又不是仅仅追求闪亮的视觉效果，不仅仅是他们身份地位的象征，也不是像真正的艺术作品那样追求完美。

设计的出发点是镶金边的卡布丽缇餐椅（Cappelletti），这几把椅子定下了整个室内空间不凡的基调。

这间公寓里有许多独特的设计亮点。比如说，公寓地板采用了一种非常罕见的昂贵木材，从里面散发着光芒，就像琥珀一样。

厨房的设计也不同寻常。设计师着手设计厨房的时候，公寓主人提出要采用一种稀有的材料，呈现出奢华大气的风格。这种关键材料就是玳瑁，并采用嵌入的黄铜来修饰。图纸也是由马利娜·普蒂洛夫斯卡亚设计工作室完成的。同样，特别的设计还有拼接天花板，采用细纹木，以镶嵌工艺制作而成。

另一设计亮点是主卧的浴室。这个小空间需要扩大其视觉效果，设计师采用的方法是利用大玫瑰花图案的大理石地面。深色的枝干和水彩的花瓣，搭配闪亮的镜面背景，营造出一种出人意料的效果，就像这间公寓里的许多空间一样，令人眼前一亮。

公寓里的每个房间都有自身的特色，但是整体又体现出统一的氛围。"同中见异"的设计是最难的，既是和谐的整体，又有不同的特色。马利娜·普蒂洛夫斯卡亚设计工作室凭借多年丰富的经验，成功做到了这一点。

5. The golden thread dining chairs Cappelletti turned out a quite unconventional interior.
6. The kitchen uses a very rare material – tortoise shell.
7. Master bedroom
8. The marble floor of the huge roses enhanced the visual effect.

5. 镶金边的卡布丽缇餐椅定下了整个室内空间不凡的基调
6. 厨房采用了一种稀有材料——玳瑁
7. 主卧
8. 超大玫瑰花图案的大理石地面凸显了空间的视觉效果

7. Satisfaction of Requirements

Home spaces, no matter in what styles, are places for living, and the ultimate goal is to satisfy residents' requirements. The interior is to be perceived with human senses, and the design should be judged by the physical and psychological perception of the user. Therefore, we may say that all design solutions for home spaces should be human-centred and the ultimate purpose is to be user-friendly. What is a user-friendly home space? What kind of a home interior can fully satisfy requirements of the user? There's no absolute answer to these questions, nor fixed standard or principles to follow. In-depth communication between the designer and client is the basis, and with human concern and the ability to implement a good design strategy, the realisation of the ultimate goal is possible.

7. 满足人的欲望

无论何种类型的家居环境，都是人所居住的环境，最终目的都是满足人的需要，由人的思想和感知来体验，家居环境设计的成功与否，最终也是要以人的生理和心理感受来判断。所以，家居中的一切设计活动都该以人为中心，把人性化作为设计的最终目的。什么样的家居环境是人性化的，什么样的家居环境能充分满足人的需求，这种问题并不能完全的量化，也没有固定的模式和守则，这一切的实现都需要设计师和业主有深入的沟通，有充分的人文关怀，以及实现这些设计蓝图的能力。

人对家居环境的需求有几个层次，首先是基本功能的满足，如睡眠，就餐等，其次是延伸功能的满足，如娱乐、学习、会客等，另外还有以人体工程学为基础的舒适性的满足，如桌子的高度，电视

Requirements concerning home spaces can be categorised into several hierarchies. Firstly there are the fundamental requirements such as sleeping and dining. Secondly come the requirement for comfort based on the human engineering design, including the height of the table, position of TV sets, and width of the passageway. Lastly we have aesthetic requirements. The fundamental requirements are necessary in every home, and design solutions are almost the same, only with differences in details. However, the extending requirements may differ greatly for different families. Reading, working, body building exercise, play... to name just a few of the possible requirements. Only through efficient communication with the client, the designer can understand the requirements of every family member, and what he should do is not only to satisfy the requirements they expressed, but also to find out potential needs of the clients that they themselves are even unaware of. Design for comfort relies on every detail. Though there are fixed dimensions recommended for home spaces, and standard heights and sizes for furniture, to make a human-centred home design you need to do more, especially for families with complicated compositions, for example, families including many members (the elder, children, and the disabled, etc.). All the details should be taken into account, such as space layout, position of furniture, height of switches, and choice of plants. The satisfaction of aesthetic requirements mainly depends on the artistic taste and abilities of the designer. Every family member has his or her understanding and requirement concerning the aesthetics of his or her home space. They might propose fragmented or vague ideas, which need to be made definite in the design process. The role of the designer is to capture the essence of their ideas and present them in an artistic way.

The client of the following project is an old couple. They wanted a warm, comfortable and artistic residence that can satisfy all their requirements. The design team perfectly satisfied their wish. In this large house with a total floor area of more than 1,000 square metres, there are a fully equipped gym, audio-visual room, study, wine cellar, play room for grandsons, library, bedroom and kitchen for servants, and outdoor swimming pool... All their requirements are satisfied, and even the wardrobes are designed with different colours according to the wearing occasions of the clothes (a problem that had long been a trouble for the couple). Everything is done in a way they prefer, without any flaw, and this is what the designers would mostly like to see.

的位置，走廊的宽窄等，最后还有审美功能的满足。基本功能是每个家居环境中必备的功能，并无太多差别，只是在每个环境的细节呈现上有不同的要求。但对延伸功能的要求每个家庭却都千差万别，阅读、工作、健身、游戏……这些功能不能全部呈现出来，只能通过充分的沟通才能了解每个居住在这个空间的人对空间的要求，设计师要做的，不仅是满足这些要求，更要在这个基础上，想到他们没有想到却依然存在的需要。舒适度的需求更多的体现在每个细节中，虽然住宅设计有许多成文的规范尺寸，家具也有标准的大小高低，但以人为本的设计往往需要更多，尤其是对一些家庭结构复杂，成员众多，有老人、儿童或者残疾人的家居环境，大到房间的布局、家具的位置，小到开关的高低，盆栽的选择，每一样都需要仔细的思考和慎重的选择。审美功能的满足在很大程度上取决于设计师的情趣和能力。每位家庭成员都会对自己的家和自己的专属领域有审美上的要求和想法，这些想法也许是零碎的、模糊的，需要在设计过程中加以明确、连贯，而设计师则需要扑捉这些零散的意识，把他们用更美的形式呈现出来。

本案例中的业主是一对老年夫妇，他们希望这所住宅能满足他们的所有需求，并且希望它温暖、舒适，有艺术性。而他们的设计团队完美的满足了他们的愿望。这所1000多平方米的大房子中有设备齐全的健身房、视听室、书房、酒窖、孙子们的游戏室、图书馆、佣人的卧室、厨房、室外的游泳池……所有他们要的功能都一一具备，就连衣橱也根据穿着场合和功能分成了不同的颜色，解决了这个困扰夫妇很长时间的问题。一切都按照他们喜欢的方式，没有任何缺点，这应该也是设计师们最乐于看到的。

1. Back view of Brown Residence
1. 布朗别墅后身

Brown Residence
布朗别墅

This house is comfort, functionality and most of all, beauty, and has left owners giddy with pleasure. It has fully equipped gym/spa, media complex, his and her studies, wine room, playrooms for the grandkids, chef-ready kitchen, guest suites, library, servants' quarters and staff kitchen, outdoor pool and indoor Roman bathroom.

The real work centred on creating the environment the owners craved. They wanted a warm, comfortable but not stodgy house; they also wanted the formality of a Great Room that looked handsome and they wanted to be able to utilise the rooms for all of their desires. When they moved into this house, it is all completed. "Everything here is just the way we wanted it, and it is like living in a resort," says the owener.

One of the happiest surprises in the new house is one that bedeviled them in the old: closets. The owner loves clothes and shopping and often picks out his wife's outfits as she would rather be on the golf course. Now the couple's closets – they are really rooms – and shelves are colour-coded, arranged by occasion and function and totally accessible.

The new abode has so many goodies and such expansive gardens that the owners, who are used to travelling far and wide, now hate to leave home. The couple is thrilled with the end result. Everything is just the way they wanted it. They use all parts of the house – even the playroom which was set up for their four grandchildren. They don't play with the giant train set, but they love the pinball and other arcade games scattered around the room.

Location Cresskill, New Jersey, USA
Designer Lauren Ostrow Interior Design, Inc.
Photographer Bruce Buck
Area 1,402m²

项目地址 美国，新泽西州，克瑞斯基尔镇
设计师 劳伦·奥斯特洛夫室内设计公司
摄影师 布鲁斯·巴克
项目面积 1,402平方米

2. Entry foyer sticks to a colour of cream.
3. Living room from second floor balcony
4. Living room from entry foyer
5. Clean and simple living room

2. 入口门厅采用统一的乳白色色调
3. 从三楼平台看客厅
4. 从入口门厅看客厅
5. 客厅简约大方

新古典家居的功能艺术 | 095

6. Mirror to reflect natural sunlight and visually create a larger space
7. Master sitting room

6. 镜面的作用是反射自然光线，并扩大空间的视觉效果
7. 主人起居室

这栋别墅集舒适、实用及——最重要的一点——美观于一体，令房主眼前一亮，非常满意。别墅内配套设施齐全，包括健身设施、SPA水疗设施、多媒体设备、男女主人各自的书房、品酒室、孙辈的游戏室、设备一应俱全的厨房、若干客房、图书室、仆人室、仆人厨房、室外泳池和室内罗马式浴室等。

本案的设计工作围绕着房主想要营造的环境而展开。布朗夫妇想要一栋看上去温暖、舒适而又不落俗套的别墅，他们想有一间"大屋"，看上去宽敞而又大气，希望别墅内的各个房间能够满足他们的所有需求。当他们搬到这栋别墅里的时候，一切都实现了！房主表示："这里的一切都正是我们想要的那样，就像每天住在度假村一样，真是棒极了。"

新别墅带来的最大惊喜之一，就是特别设置了把我们带回古老时光的衣帽间。房主喜欢服装和购物，经常需要为他的夫人选购全套装备——这位太太爱打高尔夫球。现在，这对夫妇有了专门的衣帽间——真正的衣帽间，而不是衣橱，里面有架子，采用不同的色彩来分门别类，根据服装的场合和功能来分类，取用十分方便。

新家的配备如此之好，屋外的花园美丽开阔，曾经喜爱远行旅游的布朗夫妇，现在却舍不得离开家了。两人对新家的装修结果非常满意。一切都是他们想要的那样。别墅的各个角落都令他们喜爱，他们甚至去用给4个孙辈孩子准备的游戏室！他们没玩里面的大型铁路模型，但却喜爱那里的弹球台和散布房间各处的其他大型电玩游戏。

Ground Floor Plan
一楼平面图

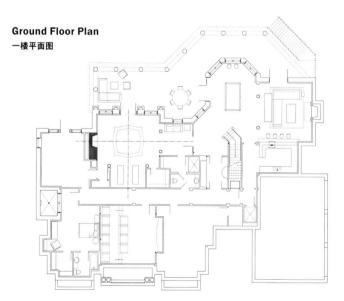

First Floor Plan
二楼平面图

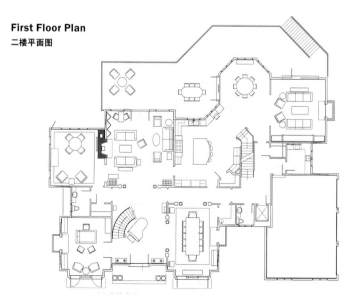

098 | FUNCTIONS OF NEO-CLASSICAL HOME DESIGN

8. Dining room from entry hall
9. Dining room full of sunlight
10. Breakfast room with terrace beyond
11. Grandchildren's bedroom
12. Library office
13. Library study

8. 从入口门厅看餐厅
9. 餐厅阳光明媚
10. 早餐室,外面是露台
11. 孙辈卧室
12. 图书室办公空间
13. 图书室阅览空间

Second Floor Furniture Plan
三楼家具布置平面图

Attic Furniture Plan
阁楼家具布置平面图

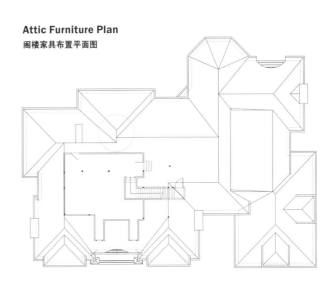

Decorations bring out visual enjoyment.

With the development of home design, we have higher demands for the aesthetics and personal characteristics of home spaces. Apart from the satisfaction of functional requirements, we want to show particular aesthetic tastes in our home, with harmonious colours, natural materials, exquisite details, and a unified style formed by these. Neo-classical style has been fully tested with time and is still a popular classic style. The unique qualities made it a timeless style.

- The way of combination of neo-classical elements and their

装饰创造视觉享受

随着家居设计发展的日益成熟，人们对家居环境的美感和个性化的追求更加强烈，在满足功能之余，人们更希望自己的家有独特的审美情趣，有精巧的造型、和谐的色彩、天然的材料，以及这些所形成的家居风格。新古典风格是经过了时间的洗礼依然受到无数人喜爱的经典风格，独特的气质赋予了它历久弥新的能力。

新古典风格元素的组合方式，造型特点，
新古典风格的细节处理方式，
新古典家居环境适合什么样的材料，
新古典风格的色彩搭配方式，

CHAPTER 3
DECORATIONS OF NEO-CLASSICAL HOME DESIGN
新古典家居的装饰艺术

characteristic shapes
- Treatments of neo-classical details
- Materials suitable for neo-classical home design
- Colour schemes for neo-classical style
- The role of furniture in neo-classical style homes
- Integration of other forms of art into neo-classical style design
- Features of neo-classical styles in different countries

All these factors have their influence on the implementation and ultimate result of neo-classical style in home spaces, and are the key points for the creation of a unique home style.

家具在新古典家居环境中的作用,
新古典风格对其他艺术形式的吸收,
不同国家的新古典风格特点。
……
这些问题关系着新古典风格在家居中的实现程度和最终的形式,也是塑造一个独一无二的家居风格的关键所在。

1. Space with Different Elements

The neo-classical style originating from the ancient Greece and Rome highlights harmony, elegance and dignity. It puts great emphasis on proportions and details. There are many typical elements that belong to the neo-classical style, but in home design, we have to be careful to choose from them.

Neo-classical furniture, classical columns, fireplaces, doors and windows with arches, crystal chandelier, oil paintings... These are some of the most commonly seen elements of neo-classical style. In home spaces, we don't need to include all; instead, we can choose according to the characteristics of the space, and the combination of the elements used should be in accordance with neo-classical design principles.

Harmony and dignity of neo-classical style design can be achieved in

1.不同元素组成的空间造型

新古典风格源于古希腊和古罗马，强调造型的和谐、优雅和庄重，对装饰造型的比例和精确度有高的要求，另外，典型的元素也有很多，但在家居空间中却需要有所选择。

新古典风格的家具、古典柱式、壁炉、拱柱结合的门窗、水晶灯、油画……这些都是常见的体现新古典风格的元素。在家居空间中，并不需要把每一样都罗列出来，而是要根据空间的特点来决定，而这些元素组合在一起呈现出来的整体造型也需要符合新古典的规则。

造型的和谐和庄重体现在很多方面，其中对称是关键的一条。例如，在较大的门洞两侧搭配对称的柱式，在壁炉装饰面的两侧雕刻对称的壁柱，或者在较大的空间内用对称的柱式分隔空间。客厅中的沙发可以采用对称的组合形式，卧室中的柜子也可以对称摆放。可以说，对称的形式是体现庄重的最重要方式。

many ways, one of which is symmetry. For example, you can place symmetrical columns on the two sides of a big doorway, add pilasters on both sides of the fireplace, or use symmetrical columns in a large room to make a vague division. Sofas can be placed in a symmetrical way; so are cabinets in the bedroom. In neo-classical style design, symmetry is the most important way to achieve a sense of dignity.

Apart from symmetry, proportion is also a big part. Proportional lines on all surfaces and the size of each panel should be carefully designed to achieve a balanced visual effect. The scale of furniture, width of doors and windows, and positions of lamps all have to be decided according to the area of the room to produce good proportions of the overall interior. In this sense, in the creation of a space, we have to understand not only its characteristics as a whole, but the qualities of all elements in it.

In neo-classical style, harmony and dignity always go well with grace and elegance, which is often achieved with textured materials, such as luxurious wallpaper. In neo-classical home design, wallpaper with patterns of vines and flowers are often adopted in a symmetrical way, producing a romantic and graceful atmosphere. Elegant crystal chandeliers are preferred because they easily create a beautiful flow of light and colour in the interior. As for the selection of textiles, lace curtains, oriental silks, Egyptian cotton and velvet are some of the best choices. They are heavy and fabulous in colour, and can be helpful in enhancing the grace and elegance of the ambiance.

In the following project, the client wanted a classic home environment, and the designers decided to combine several elements of neo-classical style in a seemingly random way. In this home design, you can find classic furniture, luxurious wallpaper, vintage oil paintings, and the marble fireplace. Details are carefully treated, with rhythmical lines and sculptural decorations. Perfect curving, delicate profiles, exquisite details and carvings can be seen everywhere, establishing an elegant taste of the timeless space. As the artists stated, the design doesn't belong to a certain style, but is a random arrangement of many elements. However, such a free combination of classical and modern elements just created a neo-classical air.

除了对称之外，比例的掌握也是造型中的一项重要内容。在各界面的装饰中，线条的比例关系，每一块镶板的面积都需要精心的设计才能达到最平衡的状态，家具的大小，门窗的宽窄，灯具的位置等也要根据室内的面积大小追求和谐的比例。所以，在塑造一个空间的过程中，不但要了解空间的特点，更要多方面的了解各个元素的性质。

新古典风格的空间造型在追求和谐庄重的同时，也有雍容典雅的一面，主要通过一些有质感的元素来体现，例如华贵的壁纸，新古典风格的壁纸常用藤蔓、花卉等图案，再加上对称的设计，给人浪漫高贵的感觉。造型典雅的水晶灯是灯具中的首选，流光溢彩的灯光效果能更衬托出环境的美轮美奂。在织物的选择上，蕾丝垂幔、东方丝绸、埃及棉、天鹅绒都是体现华丽最好选择，这些质感厚重、色泽艳丽的织物能充分地展现空间中的雍容华贵。

在本案例中，业主想要一个经典的家居环境，而设计师则选择了新古典风格中的若干元素加以随机的组合。在这个环境中，有经典的家具、华丽的壁纸、复古的油画、大理石壁炉，在细节处，也有富于韵律感的线条和富于雕塑感的装饰。无论是哪一处都有完美的弧度，精致的轮廓，讲究剪裁的精细和雕刻镶嵌的美感，处处的精雕细琢塑造了高雅的品位，也让这个空间有了穿越历史的时代感。正如设计师所说，它不属于一种特定的"风格"，只是许多元素的随机排列，但正是这种对古典和现代元素都信手拈来的排列，使空间具备了新古典的气质。

1. Ceiling detail, 170m^2 Apartment
2. Stone carving, 170m^2 Apartment
3. Carved doorways, 170m^2 Apartment

1. 170平方米公寓天花特写
2. 170平方米公寓石雕特写
3. 170平方米公寓门饰雕花特写

170m² Apartment
170平方米公寓

The owners of this flat had lived in a modern interior for a long time. Until one day they decided to change it for the classic one. The designers explain that they didn't set a task to create a project in some particular style; they were free to choose different elements and assemble them in random order. The most important thing was to create a feeling of a classical interior, of the apartment that has a history, that has been inherited by many generations for centuries. The interior of the flat reminds Old Moscow mansions and the living room is fitted with a fireplace. The building itself was constructed in the beginning of the 20th century and the fireplaces have preserved their operability since that time, having only been decorated with marble and onyx.

At the same time the interior doesn't look like a museum. To bring more modernity to it trendy colours were used. The kitchen and the bedroom are made in pearly-grey tones. The living room and the hall are golden-mustard. The colours have been brought to harmony due to what the interior looks whole. The right ratio of warm and cold tints makes it light and airy; they supplement and accentuate each other.

The study is the only bright room in the flat: the desk with a red lacquered top, a lamp with a red striped shade and crimson bookcases. Such furnishings will never keep you bored. The prevailing neutral colors tune you to relaxation, but the working place – the study – was to be rather invigorating. Hand-made elements prevail in the interior design. Besides the fireplace there are carved doorways with elaborate profiles reminding of Russian manor architecture of the 19th century. The ceiling painting where the motives of the trellis are interpreted (the owner's bedroom) arises the same associations as well as the delicate floral patterns on the glass.

Location Moscow, Russia
Designer Anna Kulikova & Pavel Mironov
Photographer Alexander Kamachkin
Area 170m²

项目地址 俄罗斯，莫斯科
设计师 安娜·库里克娃与帕维尔·米若诺夫
摄影师 亚历山大·卡玛奇金
项目面积 170平方米

4. The bedroom has a feeling of a classical interior.
5. Entrance hall
6. The hall is golden-mustard.
7. The dining room is made in pearly-grey tones.

4. 卧室凸显古典风格
5. 入口门厅
6. 门厅采用金黄色色调
7. 餐厅的色调采用珍珠灰

这套公寓的业主曾居住在一个现代风格的室内空间里，有一天他们决定要将其改变成古典的风格。设计师们解释说他们并没有将其设定成一个具有特殊风格的项目；他们很自由地选择不同的元素，并且将它们随机组合到一起。最重要的是要创造出一种古典的室内氛围，使这套公寓散发一种有着几百年传承的历史感。

这套公寓的室内风格会令人联想到莫斯科古老的宅邸，起居室配置有壁炉。这座建筑本身建于20世纪初期，壁炉保持了那一时期的风格，仅以大理石和玛瑙装饰。同时又不能使室内看起来像一间博物馆，所以设计师选用了时尚的色彩使其更具有现代感。厨房和卧室的主色调为珍珠灰，起居室和大厅的主色调为金黄色。各种色彩的运用使室内更加和谐统一。冷暖色调的合理搭配相得益彰，使室内空间更加明亮而具有通风感。

书房是这套公寓中唯一一间仅使用了明亮色彩的房间：红漆面的办公桌、红条纹的灯具，还有深红色的书架。这样的家具陈设不会令您感到厌烦。流行的中性色可以帮您调整心情，彻底放松，但是，工作场所——书房的设计要能够令人精神振奋。室内设计中充满了手工制作的元素。壁炉旁的门廊上方精心雕刻的浮雕令人联想到19世纪俄罗斯的庄园建筑。主卧室天花板上的绘画诠释了天花板框架结构设计的动机，并且与玻璃上精美的花形图案相互衬托，相互辉映。

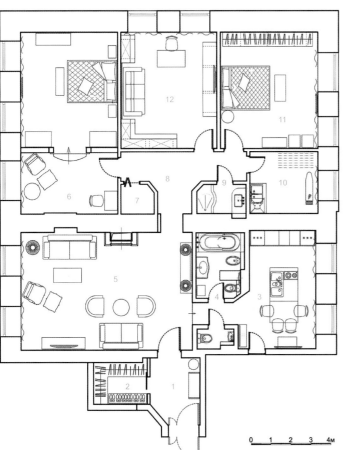

Floor Plan
1. Entrance hall
2. Dressing room
3. Kitchen
4. Bathroom
5. Living room
6. Master bedroom
7. Storage room
8. Corridor
9. Shower room
10. Laundry
11. Bedroom
12. Office

8. A fireplace is the centre of the living room.
9. The colours have been brought to harmony in the living room.
10. The right ratio of warm and cold tints makes it light and airy.
11. The bedroom has different elements.
12. A corner of bedroom
13. The bathroom has the same tone with the bedroom.
14. Master bathroom
15. The owner's bedroom
16. The wall in the bedroom has delicate floral patterns.

平面图
1. 入口大厅
2. 更衣室
3. 厨房
4. 卫生间
5. 客厅
6. 主卧室
7. 储藏室
8. 走廊
9. 淋浴间
10. 洗衣房
11. 卧室
12. 办公室

8. 壁炉是起居室的核心元素
9. 起居室各种色彩的运用非常和谐
10. 冷暖色调的合理搭配相得益彰，使室内空间更加明亮而具有通风感
11. 卧室采用多种装饰元素
12. 卧室一角
13. 浴室采用和卧室相同的色调
14. 主浴室
15. 主卧室
16. 卧室墙面有精致的花纹

2. Dynamic and Luxurious Details

The success of any interior design, no matter what styles it belongs to, lies in careful treatments of details. Particularly in home spaces, details would attract more attention. Stepping from the complicated Rococo style, neo-classicism has been developed to present an implicit magnificence, and delicate and creative details are always needed.

Concerning decoration, line carvings are usually preferred in neo-classicism to cope with other elements in the space. The carvings may appear on door frames, mouldings where the ceiling and the wall meets, and on handrails, etc., with sculptural patterns inspired by nature, including vines, flowers and animals. They can coexist well with furniture, wall carvings, lamps and other elements in the space, completing both a beautiful environment and graceful details.

2. 细节处的生动和华丽

一种风格的成功塑造离不开对细节的关注。尤其是在家居这个环境中，细节往往会被更多地注意到。新古典风格从繁复装饰的洛可可时代走出来，想要达到更加含蓄高雅的效果，就需要更精确更有灵性的创造。

在装饰上，新古典风格常用细节处的线条雕刻搭配空间中的其他元素，这些雕刻会出现在例如门框上、天棚和墙面的线脚处、栏杆上等各种地方，雕工细致，有很强的立体感，花纹一般都来源于自然，藤蔓、花叶、动物图案等，并且与空间中的家具、墙面雕刻、灯饰等形成一定的呼应，使空间无论从整体还是局部都有优美的曲线。在棚顶或地面也会做出一定的造型，圆形、椭圆形、波浪形，并且地面和棚顶的造型往往上下一致。墙面有时也会增加拱形的造型与拱形门和拱形窗户呼应。除此之外，一些小型的饰品也会为空

Floors and ceilings can be treated with details, too. Circular, oval, or wavy carving patterns can be added. Usually the ceiling and floor surfaces would present consistent designs. We can add arches on the wall to correspond with arch doors or windows. In addition, small ornaments can be helpful, and are often preferred on the bedside, on the fireplace, or on the tea table. Selection of the ornaments should be careful to match with the surrounding environment, and the ornaments themselves should be distinctive. In a neo-classical home, we often find gilded candlesticks, angel sculptures and other pieces of artwork placed on the fireplace, lambskin or lace lampshades on the bedside table, metal or natural stone lamp holders, vintage bronze mirrors, sculptures or photos on the low cabinet or tea table… Details of the textile are often ignored. In neo-classical home design, heavy and gorgeous textiles are preferred, but here ornaments can also find their way. For example, you can have embroidered edging or tassels on the window curtain, patterns on the cushion, textures on the carpet, etc. All these details would affect the coordination of the whole space. Textures and patterns of the textile should be harmonious with that of the surroundings.

In the following project, the designers paid great attention to the details. This is a residence like an Italian palace. A spiral staircase is placed at the entrance, whose metal handrails are designed with natural patterns, corresponding with the huge chandelier on the ceiling and distant guardrails of the window. Railings at other parts of the house are designed with the same patterns. The staircase features hand-carved floral patterns under the handrails, waving upwards to the upper floor, further enriching the stairs. Ceilings in the house are another highlight. The ceiling in each room is carefully carved and designed with an oval shape, varying in size according to the area of the room, being simple or complicated depending on the function of each space, but all with beautiful curving lines and balanced proportions, coexisting well with the surrounding arch doors or windows. Curves serve as a prevailing theme of the design. Apart from arch doors and windows, arches can be found in many parts of the interior, and there is curvy furniture as well. Circular patterns appear on the large-span glass on the façade. All in all, no matter inside or outside, curves play an important role in the design of the residence, and harmony of neo-classicism is successfully achieved.

间增色。在床头、壁炉架上，或是边桌上，都需要一些饰品点缀，饰品的选择要与环境搭配和谐，同时也要有一定的特点。在新古典风格中，壁炉架上常会有镀金的烛台、天使造型的雕塑，或其他工艺品，床头摆放羊皮或带有蕾丝花边的灯罩，铁艺或天然石磨制的灯座，矮柜或茶几上装饰复古的铜镜、雕塑或照片等。细节处理中织物的处理常常会被忽略，新古典风格的织物选择往往以华丽大气为主，但在织物上也需要有一些装饰，例如窗帘上的镶边，垂饰、靠垫的花纹，地毯的材质，这些细节的选择也会影响环境的协调性，所以，选择质地和花纹同环境和谐的织物同样重要。

在本案例的内部设计中，设计师对细节的装饰给予了很大关注。这是一处意大利宫殿般的住宅，入口处有一个旋转的楼梯，楼梯上的铁艺扶手被设计成了极富自然情趣的花纹，与顶棚上的大型吊灯和远处窗户边的护栏交相辉映，而其他各处的栏杆也都采用了这种花纹。楼梯最有特色的是在铁艺栏杆下面还有手工雕刻的花朵图案，一直蜿蜒到上层，使楼梯变得更加丰富。另一个值得注意的细节是住宅中的天棚，每一个房间的天棚都经过了精心雕刻，并都采用了椭圆形的造型，这些椭圆的造型根据房间的大小和功能或简单或复杂，但都有着优美的曲线和适当的比例，与周围拱形的门窗共同形成了完美的弧度。曲线在这个空间中的运用很多，除了拱形的门窗，室内还有带有弧线的家具和拱形的造型，建筑外立面的大型玻璃上也有圆形的装饰，无论是室内和室外，都能看出曲线对这所住宅的影响，这也是对新古典风格所强调的和谐的一种印证。

1. Exterior of 3rd Avenue

1. 第三大道住宅外景

3rd Avenue
第三大道住宅

One of the requests by the clients was that Nico van der Meulen Architects builds them an Italian palace. The house has eight large bedrooms, each with their own en-suite bathrooms, including a guest room on the ground floor. The main bedroom has its own private lounge as well as his and her dressing room. For every two children's bedrooms there is a lounge attached.

There is a formal lounge and dining room that both have mezzanine balconies within. These balconies can be used for entertainment or speeches.

The family room and lanai overlooks the indoor pool, jacuzzi and water feature.

The dramatic entrance features a winding staircase with an ornate balustrade and hand-carved patterns along the side of the staircase.

There are three atriums situated one on each floor of the house. The basement level accommodates a billiard room, a 10m pool with an automatic pool cover, a home theatre, jacuzzi, Japanese eating area at the pool, water feature, gym, massage room, sauna and steam room. The ceiling height in the pool area is approximately four stories, creating a light airy space.

The gazebo can seat sixty people, and has its own kitchen and male and female bathrooms. Included on the site are staff cottages and a guard house. The total area of the house is 2,500m² and the staff buildings total 250m². The garage can house eight cars.

The house is situated on dolomite rock so the house was designed on a waffle foundation, constructed at a 1.6m depth to withstand a 25m diameter sinkhole to prevent cracking. The site falls from west to east approximately two stories.

The gilding and painting on the ceilings were hand crafted and includes a hand-painted fresco of the 99 names of Allah in the prayer room. The tower in the prayer room is almost four stories high and the ceiling was hand-painted over several weeks. This dramatic Italian home exudes drama, opulence and attention to detail.

Location Pretoria, South Africa
Designer Nico van der Meulen Architects
Photographer Nico van der Meulen Architects
Area 2,500m²

项目地址：南非，比勒陀利亚
设计师 尼科范德莫伊伦建筑师事务所
摄影师 尼科范德莫伊伦建筑师事务所
项目面积 2,500平方米

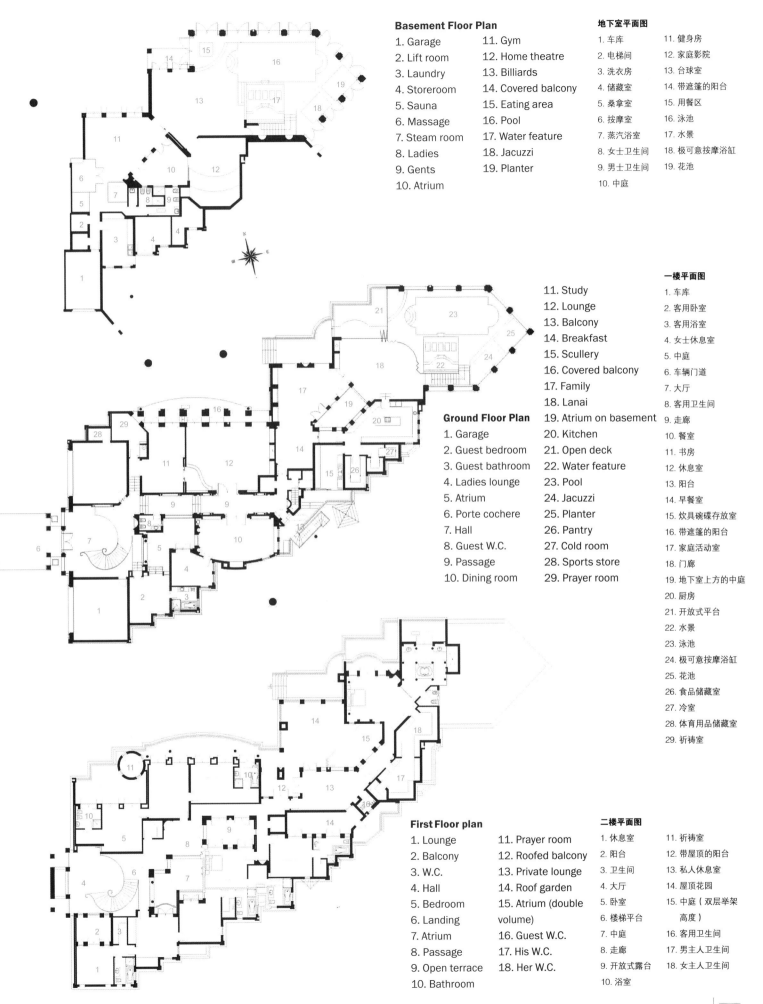

Basement Floor Plan

1. Garage
2. Lift room
3. Laundry
4. Storeroom
5. Sauna
6. Massage
7. Steam room
8. Ladies
9. Gents
10. Atrium
11. Gym
12. Home theatre
13. Billiards
14. Covered balcony
15. Eating area
16. Pool
17. Water feature
18. Jacuzzi
19. Planter

地下室平面图

1. 车库
2. 电梯间
3. 洗衣房
4. 储藏室
5. 桑拿室
6. 按摩室
7. 蒸汽浴室
8. 女士卫生间
9. 男士卫生间
10. 中庭
11. 健身房
12. 家庭影院
13. 台球室
14. 带遮篷的阳台
15. 用餐区
16. 泳池
17. 水景
18. 极可意按摩浴缸
19. 花池

Ground Floor Plan

1. Garage
2. Guest bedroom
3. Guest bathroom
4. Ladies lounge
5. Atrium
6. Porte cochere
7. Hall
8. Guest W.C.
9. Passage
10. Dining room
11. Study
12. Lounge
13. Balcony
14. Breakfast
15. Scullery
16. Covered balcony
17. Family
18. Lanai
19. Atrium on basement
20. Kitchen
21. Open deck
22. Water feature
23. Pool
24. Jacuzzi
25. Planter
26. Pantry
27. Cold room
28. Sports store
29. Prayer room

一楼平面图

1. 车库
2. 客用卧室
3. 客用浴室
4. 女士休息室
5. 中庭
6. 车辆门道
7. 大厅
8. 客用卫生间
9. 走廊
10. 餐室
11. 书房
12. 休息室
13. 阳台
14. 早餐室
15. 炊具碗碟存放室
16. 带遮篷的阳台
17. 家庭活动室
18. 门廊
19. 地下室上方的中庭
20. 厨房
21. 开放式平台
22. 水景
23. 泳池
24. 极可意按摩浴缸
25. 花池
26. 食品储藏室
27. 冷室
28. 体育用品储藏室
29. 祈祷室

First Floor plan

1. Lounge
2. Balcony
3. W.C.
4. Hall
5. Bedroom
6. Landing
7. Atrium
8. Passage
9. Open terrace
10. Bathroom
11. Prayer room
12. Roofed balcony
13. Private lounge
14. Roof garden
15. Atrium (double volume)
16. Guest W.C.
17. His W.C.
18. Her W.C.

二楼平面图

1. 休息室
2. 阳台
3. 卫生间
4. 大厅
5. 卧室
6. 楼梯平台
7. 中庭
8. 走廊
9. 开放式露台
10. 浴室
11. 祈祷室
12. 带屋顶的阳台
13. 私人休息室
14. 屋顶花园
15. 中庭（双层举架高度）
16. 客用卫生间
17. 男主人卫生间
18. 女主人卫生间

新古典家居的装饰艺术

应客户的要求，尼科范德莫伊伦建筑师事务所将其打造成一座意式宫殿。这套住宅设有8间大型的卧室（包括一间位于一层的客房），每间卧室都设有独立的套房浴室。主卧室配有私人休息室和男女更衣室。每两间儿童卧室共享一间休息室。

该住宅设有一间正式的休息室和餐厅，都配有夹层阳台。阳台可以被用做休闲娱乐或者演讲致辞之所。

从家庭套房及其阳台中可以俯瞰室内泳池、极可意水流按摩浴缸和水景区。

入口处的盘梯非常引人注目，盘梯两侧搭配着装饰栏杆和手工雕刻的图案。

房屋三层楼中的每一层都设有一个中庭。地下室层设有一间台球室、一个长10米并配有自动泳池盖的游泳池，一间家庭影院，极可意水流按摩浴缸、池畔日式餐饮区、水景区、健身房、按摩室，还有桑拿和蒸汽室。泳池区的天花板高度大概有四层楼高，创造了一个明亮而又通风的空间。

露台上可以容纳六个座椅，配有独立的厨房和男女浴室。该场址上还建有员工宿舍和一间警卫室。房屋的总面积达2500平方米，其中员工建筑所占面积为250平方米。车库可以容纳6台车。

由于房屋位于白云灰岩上，所以将房屋的地基设计成了华夫网状，建造深度达1.6米，这样可以防止直径为25米的污水坑破裂。该场址从西到东大约有两层楼的距离。

房屋的室内是由MSQUARE LIFESTYLE DESIGN担当设计的。天花板上的镀金和绘画都是手工精心制作的，还有祈祷室中的手绘壁画，上面绘有阿拉的99个尊名。祈祷室中的塔楼大约有四层楼高，天花板上的手绘历时数周完成。这座意大利风格的住宅处处引人注目，散发着戏剧艺术风采，彰显着富足，以及对每个细节的重视。

2. Hand-carved patterns along the side of the staircase
3. Lounge on the ground floor
4. Lounge full of sunlight
5. Dining room can be used for parties.
6. Spacious kitchen
7. Interior view of pool area when pool covered
8. Interior view of pool area when pool uncovered
9. Bedroom for guests
10. Luxury bathroom
11. Dramatic Italian bedroom

2. 楼梯侧面有精致的手工雕花
3. 一楼休闲区
4. 休闲区阳光明媚
5. 餐厅里可以开派对
6. 宽敞的厨房
7. 室内泳池（封闭）
8. 室内泳池（开启）
9. 客房
10. 奢华浴室
11. 意大利风情卧室

3. Mix and Match of Materials

The selection of materials is an important part in interior design, especially in home design. Besides concerns for aesthetics, materials are a decisive factor for the realisation of sustainability, functionality and comfort of the interior space. For neo-classical style design, a good selection of materials is even more important because the magnificence of neo-classicism calls for a high requirement for materials. Therefore, as a designer, you should have an in-depth understanding of various materials suitable for neo-classical style and the ability to use them flexibly.

Neo-classicism has been in existence for hundreds of years. Now we have numerous new types of materials available. Other than traditional materials such as wood, stone, metal and ceramics, we

3. 多种材料之间的混搭艺术

材料的选择是设计中的重要环节，尤其是居住空间，除了美观之外，还关系着空间的环保、实用性、舒适性。对新古典风格来说，材料的选择更为重要，华贵的气质使其对材料的品质要求极高，因此，也需要设计师对各种适合这种风格的材料有深入的了解并能灵活的应用。

新古典风格已经产生了几百年，室内设计的材料也有了新的类型，除了传统的木材、石材、金属、陶瓷之外，还有塑料、玻璃、合成金属等，这些材料在新古典风格中都能得到合理应用。

木材可以说是在新古典风格中应用最广泛的材料。从室内的地板、到墙面的护墙板，到家具，木材都是不可缺少的材料，木材无污染，热导率小，而且能带来温暖的感觉，所以，常常会用在卧室等

have plastics, glass and synthesis metal, to name just a few. All these materials can be applied to neo-classical home design.

Wood is the most commonly used material in neo-classical style design. From floors and wall panels to furniture, wood is absolutely indispensable. It is pollution-free, has a good thermal performance, and brings out a warm feeling in the interior. Therefore, it is often applied to floors of bedrooms and other parts of the house. Besides, wood is good at heat preservation and has beautiful natural textures, making it a good choice for wall panels in neo-classical home design. However, natural wood can be very expensive, especially some rare species, so more often than not, large areas of wood would be replaced with composite wood.

Marble is one of the most frequently applied stone materials in neo-classical home design. It is solid, crushing-resistant, easy to clean, has good anti-corrosion properties and diverse patterns. Light coloured marble contributes to a sense of elegance and dignity, while dark marble feels luxurious. Both are suitable for neo-classical style, and are frequently used on interior surfaces. In home spaces, marble is mainly used as floor of the public area. Different patterns are available to go with the characteristics of spaces. Horizontal patterns would be helpful in expanding the visual width of the space, while small patterns can make a space feel larger.

As for metal materials, iron is often used in small scales, such as for handrails, little ornaments, or chandeliers. On furniture you can find bronze handles or small decorations. Metal frames can be added to photos and mirrors to add a three-dimensional effect. While metals are only used in small parts, various kinds of textiles can be applied in large areas, such as silk or velvet curtains, wool carpets and different kinds of cotton and linen materials. Though gorgeously textured textiles are preferred for neo-classical design, for furniture and beddings that are in frequent physical contact with the user, we'd better choose soft ones, with good qualities and patterns that match up with the neo-classical style.

In the following project, the selection of materials is particularly bold. Apart from traditional wooden floor, natural stone in the kitchen and ceramics in the bathroom, there are metal chandeliers decorated with quartz and crystal, bronze lamp holders, and wooden furniture and rattan chairs. As for textiles, textured velvet and linen materials are used, in addition to custom silk curtains. The mix and match of these materials creates a quiet and comfortable home environment, as well as a unique home style.

区域的地面上，木材的保温调湿功能和天然优美的纹理也使它成为新古典风格中常用的护墙板材料。但是，天然的实木造价昂贵，尤其是一些珍贵的木材越来越稀少，所以，很多情况下，大面积的木材使用会用复合木材替代。

在石材中，大理石是新古典风格常用的一种，大理石结构致密，抗压性强，耐腐蚀，易清洁，并且纹理多样，浅色的大理石装饰效果庄重高雅，深色的大理石装饰效果华丽高贵，无论哪一种都与新古典风格的要求不谋而合，所以经常会用在室内界面的装饰上，而在家居空间中，一般会在公共区域的地面上，根据空间的特点选择不同的纹样，水平的纹理可以增加空间视觉宽度，而小图案的纹理会使人觉得空间更大。

在金属材料中，铁艺常会被用做小部分的装饰，例如栏杆处的扶手，小的装饰品，或者是吊灯，而在家具上也常常会有铜制的把手或小装饰，照片、镜子也会镶有金属框增加立体感。金属的应用在室内一般只是小部分的点缀而很少大范围使用。相比之下，各种类型的织物是家居中需要大面积使用的，例如丝绸或天鹅绒的窗帘，羊毛地毯，以及各种棉麻制品。对新古典风格来说，需要质感华丽的织物，但对那些需要经常与人接触的家具或床品上，织物的选择更以舒适为主，但是依然要有良好的材质和符合新古典风格的图案。

本案例中的材料选择很大胆，除了传统的木制地板、天然石材装饰的厨房、镶有瓷砖的浴室，还有加入了石英和水晶石的铁艺吊灯，铜制的灯座，家具除了木制以为，还有用于休闲的藤椅。织物中加入了有质感的天鹅绒和亚麻纺织品，还有定制的丝绸窗帘。这些材料混搭在一起，创造了一个舒适安静的环境，也塑造了独一无二的风格。

1. Patio in Valley Falls Estate

1. 峡谷瀑布别墅庭院

Valley Falls Estate
峡谷瀑布别墅

This home was designed with a well-curated combination of colours, materials and textures. The goal was to appear as if it had evolved over time, achieved by mixing the rustic with elegant, antique with new, and even combining contemporary art with antiques. Texture was extremely important. The designers were not afraid to mix iron, wood and stone, and add depth with textural textiles like velvet and linen.

Arches are a recurring theme throughout the home, as are the 3m high solid Mahogany doors manufactured in Honduras. Cast stone fireplaces, plastered, glazed walls and reclaimed wood beams lend themselves to both an elegant and comfortable design.

In the kitchen, a terracotta relief worked into the backsplash sets the tone along with Texas limestone walls and open reclaimed wood shelving.

A working pantry or service kitchen behind the main kitchen is outfitted with not only additional workspace, but an extra dishwasher, sink, refrigeration and covered dry pantry storage area as well. The black and crème Walker Zanger tile and custom iron pot rack add additional ambiance and function to the space.

The kitchen is open to the family room with casual dining in between. Corner bookcases on either side of the fireplace are made cosy with built-in window seats.

Custom silk drapes adorn French doors in the dining room. An iron chandelier with rock crystal, quartz and topaz hangs from a distressed beam over the glass top table.

Three-metre-high Mahogony doors lead from the formal living room to the covered veranda with fireplace. The European-inspired back yard was achieved with repeated use of boxwoods and Japanese maples, rosemary and Angelonia planted in the beds. A custom fountain on the limestone wall with creeping Boston ivy completes the plan.

Floor-to-ceiling velvet draperies adding texture to the master bedroom, a French headboard upholstered with an antique needle point rug, custom bedding including a foot warmer and commissioned hand-painted chests are just a few of the special touches. Antique French chairs and bench complete the room.

The soothing master bathroom features an antique terracotta relief set in mosaic tiles as well as a vintage tub. The dressing area is outfitted with a kidney-shaped antique vanity, fabulous sconces and a 19th-century gold gilt mirror.

Custom headboard and bedding flanked with concrete side tables create a tranquil setting in the guest room. Lucite lamps and starburst mirors add the perfect touch.

Location Little Rock, Arkansas, USA
Designer Firm Providence Ltd. Design / Mona Thompson and Talena Ray
Photographer Nancy Nolan
Area 574m²

项目地址 美国，阿肯色州，小石城
设计师 弗尔姆·普罗维登斯设计公司（莫纳·汤普森、塔列纳·雷）
摄影师 南希·诺兰
项目面积 574平方米

这栋别墅的设计巧妙地融合了色彩、材料和质地。设计目标是使其呈现出饱经风霜的感觉，这一点通过粗犷加典雅、古老加现代的混合手法得以实现，甚至将现代艺术与古董相结合。质地感非常重要。设计师大胆地将铸铁、木材、石材等质地各异的材料相结合，并利用织物（如天鹅绒和亚麻布等）特有的质感来增加空间的深度。

拱门是这栋别墅贯穿始终的一大特色，采用坚固的红木大门，高3米，产自洪都拉斯。铸石壁炉（外层抹上灰泥）、玻璃隔断和回收利用的木头横梁，共同营造出舒适、典雅的室内空间。

厨房炉灶后挡板上有赤陶浮雕，确立了空间的氛围基调。墙壁采用的石灰岩产自得克萨斯州。此外，还有开放式置物架，采用回收利用的木材制成。

主厨房后面是餐具室或者备餐室。这里不仅有额外的备餐空间，而且还有另外的洗碗机、水槽、冷柜和餐具储藏区。黑色和乳白色的"步行者尚格"（Walker Zanger）品牌瓷砖以及定制的铁质锅具悬挂架，不仅完善了备餐室的功能，而且赋予这一空间别致的氛围。

厨房与家庭娱乐室相连，中间是非正式用餐空间。壁炉两边的角落里各有一个书橱，嵌入式的靠窗座位让空间更加舒适。

饭厅里，定制的丝质窗帘装饰着法式落地双扇玻璃门。铁质枝形吊灯采用水景、石英和黄宝石装饰，从一根古旧的横梁上垂下来，悬在玻璃桌面上方。

3米高的红木大门将人从起居室引至阳台。阳台不露天，里面有壁炉。后院大量使用黄杨木，花池里种植了鸡爪枫（日本枫树）、迷迭香和天使花，营造出浓郁的欧洲风情。石灰岩墙边是为本案量身定制的喷泉，旁边爬满波士顿常春藤。

天鹅绒落地窗帘为主卧增添了质感。法式床头板采用针绣毯装饰，古韵十足。寝具都是定制的，包括脚炉。衣柜上是手绘的花纹。主卧里像这样的精致设计比比皆是。此外还有古色古香的法式座椅和长凳。

主浴室拥有抚慰人心的氛围。镶嵌的瓷砖墙面上是凸显古风的赤陶浮雕，浴缸也是古式风格。浴室更衣间内安装了古色古香的肾形梳妆台，还有考究的壁灯和19世纪的金边镜子。

客用卧室里有定制的床头板和寝具，搭配别致的床头桌，营造出舒适温馨的氛围。有机玻璃灯具和星光镜让房间更加耀眼。

2. Three-metre-high Mahogony doors lead from living room to the covered veranda.
3. Covered veranda
4. Custom silk drapes adorn French doors in the dining room.
5. The kitchen is open to the family room with casual dining in between.
6. Working pantry or service kitchen behind the main kitchen
7. Study
8. Kitchen with black crème Walker Zanger tile
9. The living room mixing the rustic with elegant

2. 3米高的红木大门连接着起居室和阳台
3. 阳台有遮棚
4. 饭厅里，定制的丝质窗帘装饰着法式落地双扇玻璃门
5. 厨房与家庭娱乐室相连，中间是非正式用餐空间
6. 主厨房后面是餐具室或者备餐室
7. 书房
8. 厨房采用黑色和乳白色的"步行者尚格"品牌瓷砖
9. 起居室融合了粗犷与典雅

Ground Floor Plan
一楼平面图

First Floor Plan
二楼平面图

134　DECORATIONS OF NEO-CLASSICAL HOME DESIGN

10. Floor-to-ceiling velvet draperies add texture to the master bedroom.
11. Dressing area in master bathroom
12. Vintage tub in master bathroom
13. Soothing master bathroom
14. Guest room with a tranquil setting

10. 天鹅绒落地窗帘为主卧增添了质感
11. 主浴室更衣间
12. 主浴室里古式风格的浴缸
13. 主浴室拥有抚慰人心的氛围
14. 客房舒适温馨

4. Diversified Colour Schemes

Colour is one of the most significant elements that will influence the visual effects and ambiance of the interior. Appropriate colour schemes would not only beautify your home spaces, but influence your mood. Therefore, colour design in home spaces should be based on aesthetic principles as well as criteria of environmental psychology. The ultimate goal is to create a comfortable, safe, and aesthetical home environment.

Neo-classical colour schemes are usually graceful and harmonious, with the whole space feeling luxurious and spacious. In design practice, there's not specific requirement about what colours are to be used, but a harmonious ensemble should be achieved by your colour scheme. In neo-classical home design, the most frequently used main colours are white, dark red and yellow, while gold is often partially used on lines and as embellishment to enhance luxury of the space. You can choose various colours according to different programmes. In the large area of the living room, neutral colours are recommended, such as ivory and light brown, and bright colours that go well with the main tone can be added to enrich the space. Multiple choices of colours would likely satisfy aesthetical requirements of different family members, and the large room would be better balanced. If it is a small living room, you may choose simple and elegant colour schemes, filling the small space with fun and interest. The dining room and kitchen are better to be visually enlarged by the use of bright colours. Soft yellow and other warm tones are recommended, which can help establish an intimate dining atmosphere. As for the private bedroom, since sleeping and resting take place here, the colour scheme should be quiet and mild; pure and peaceful colour schemes are recommended. In addition, attention should be paid to the coordination between the colour scheme and the textiles used, whose textures and patterns are better to be integrated with the bedroom colour schemes.

In home design, there is no fixed standard for the choice of colours, but you should have primary and secondary colours and make them combined into a harmonious whole. In neo-classical home design, colours like blue, green and purple can be good choices for secondary colours.

In the design of the following project, the colour scheme is a focus. Different from the most commonly seen dark and dignified tone in

4.多重色彩的装点

色彩是影响室内环境视觉效果最重要的要素，也是影响家居氛围的重要条件。适合的色彩不仅能使家居环境变得更美观，也能影响居住其中的人的情绪。所以，在家居空间中的色彩设计不仅要有美学的依据，同样也要依照环境心理学的规律，创造出舒适、安全并富有美感的环境。

新古典风格的色彩搭配要求高雅、和谐，使空间给人华贵、宽敞的感觉。在实际应用中，并不拘泥于哪几种色彩，而是要达到整体效果的和谐。常见的主色调有白色、暗红色、黄色，金色常用在部分线条的勾勒和局部的点缀以体现奢华的氛围。尽管如此，家居空间中也可以根据功能区的不同选择合适的色彩。在大面积的客厅中可以使用比较中性的色调，例如乳白色、浅褐色，并搭配同类的更明

neo-classical design, in this house, many bright and bold colours are used, including burgundy, green, black, blue and purple. Besides, each room is unique, differing from others in terms of colour scheme. At the entrance hall, soft, light yellow is well matched up with dark red, creating a warm and intimate welcoming ambiance. The living room is mainly white, with gold partially used, as well as purple sofas, which centrally dominate the entire living room and wipe out the dull and depressing feeling that otherwise may arise in the space. In the master bedroom, neutral colours are used as the main tone, with some golden inlays. Textiles with large areas of floral patterns are used to enliven the space. In the children's room, a lively mix and match of blue and white is created, with interesting patterns appealing to children, feeling full of imagination. The most distinctive space in the house is the bathroom for the mistress. Totally different from the conventional simple colour schemes in bathrooms, this space is immersed in multiple strong colours such as blue, green and purple, creating intense visual impacts. These colours appear on the floor, walls, curtains and mirrors, and the whole space is brightly enlivened. It is a new and successful try for innovation in neo-classical home design.

快鲜艳的色系以增加空间的层次。这样的选择可以满足更多人的审美意愿，也使大空间显得更庄重。如果是面积较小的客厅，还可以选择更加淡雅清新的色调，让小空间更有情趣。餐厅和厨房需要明快的色彩加大空间的视觉效果，可以选择柔和的黄色或其他暖色调增强就餐时的温馨气氛。而卧室的色彩应该以宁静、柔和为主，卧室的私密性很强，并且主要用于睡眠和休息，所以应该尽量更纯粹，安静的色彩组合，并且要注重色彩与空间中织物的搭配，使织物的质地、花色能和卧室中的色彩相互融合。

在家居空间中，色彩的选择并没有一成不变的规定，但要有一定的主次，在和谐统一中有层次和变化，新古典风格中，蓝色、绿色、紫色等都可以用做配色。

本案例中的色彩搭配是设计中最有亮点的部分，不同于新古典风格中常见的稳重深沉，在这所住宅中，同时运用了许多大胆明亮的色彩，酒红色、绿色、黑色、蓝色、紫色……并且，每个空间都不尽相同，有自己的特点，入口大厅处的柔和的浅黄色搭配暗红色，舒适温馨，客厅采用了白色和部分金色，并搭配了紫色的沙发，这样的配色使位于客厅中心的沙发在空间中有了更强的存在感，也打破了大部分白色和金色的沉闷。主人的卧室同样是中性色加入少量金色的镶嵌，并且用了大面积的花朵纹样的织物来使空间更活泼。而在儿童房中，充满了大量的蓝色和白色的搭配，并绘有极富童趣的图案，使空间欢快活泼，充满想象力。最有特色的是为女主人设计的浴室，一反浴室色调简单的常态，用了大量蓝色、绿色、紫色等强烈的颜色创造了一个让人目不暇接的浴室空间，这些浓重的色彩出现在地面、墙面、窗帘、镜子等任何一处，使空间中每一个部分都充满了生机，也为新古典风格找到了一个新的方向。

1. Hall in Romanovo house

1. 罗曼诺夫别墅大厅

Romanovo
罗曼诺夫别墅

This house was built for a beautiful woman, a businesswoman with a bright personality.

In this interior were mixed the principles of classical and modern techniques. In this house, every room has its own character and mood. Bold and bright colours (green, red, burgundy, black, blue, and violet) give each room uniqueness and idleness. All the sketches were drawn by Marina Putilovskaya. Each fragment of stucco, each design of a stone or parquet floor mosaic is unique and was made solely for this project.

In this house was everything – high ceilings, spacious rooms, large windows.

On the ground floor public areas are located – a large hall, dining room, main hall with a large fireplace and a terrace facing the windows.

On the second floor are owners' rooms, provided with two bathrooms and dressing rooms, and children's rooms. In addition to bathrooms and closets for children was also provided a large game room.

The house also has two kitchens, one for presentation, which plays the role of the home bar, and the second is a professional one.

On the first floor the designers added a spacious relaxation area with a swimming pool, sauna and spa, from where you can enter the terrace and the garden. They also replanned the entrance area and changed the shape of the ladder, creating a beautiful central hall.

In this house Design Bureau of Marina Putilovskaya kept their favourite idea of animalistic designs. The morning sky in the swimming pool, the image of birds of paradise, exotic flowers – all these make you feel young and happy... And the interior of the house was made for a long and lucky life.

Location Moscow, Russia
Designer Design Bureau of Marina Putilovskaya
Photographer Zinur Razutdinov
Area 2,000m²

项目地址 俄罗斯，莫斯科
设计师 玛丽娜·普蒂洛夫斯卡亚设计工作室
摄影师 金努尔·拉祖迪诺夫
项目面积 2,000平方米

2. Professional kitchen in the house
3. Spacious entrance hall
4. Living room on the ground floor
5. Luxury dining room

2. 别墅里有专业的厨房
3. 入口大厅宽敞明亮
4. 一楼起居室
5. 奢华饭厅

这栋别墅是为一位美丽的女士而建的，一位个性爽朗的商界女性。

本案的室内空间融合了古典设计风格和现代技术。这栋别墅内的每个房间都有自己的特点和氛围。明亮、大胆的色彩（包括绿色、红色、紫红色、黑色、蓝色和紫色等）让每个房间显得独特而慵懒。所有的草图都是由马利娜·普蒂洛夫斯卡亚亲手所画。各个部分的灰泥以及所有的石材和镶花地板的设计，全都是独一无二的，为本案专门设计的。

这栋别墅的设计非常大气：高高的天花板、宽敞的空间、大体量的开窗。一楼是公共空间，有宽敞的大厅、饭厅、主厅（里面有个大型壁炉）和朝向窗户的平台等。三楼是卧室，包括两间浴室、更衣室以及儿童房。为儿童准备的除了专用浴室和壁橱之外，还有一间宽敞的游戏室。

这栋别墅还有两间厨房。一间有着亮丽的外观，主要用作家庭酒吧；另外一间主要用于烹调，设备齐全。

设计师在二楼增加了一个宽敞的休闲娱乐区，有泳池、桑拿和SPA水疗设施，从这里可以进入平台和花园。入口空间也重新进行了规划，改变了楼梯的造型，营造了一个美丽的中央大厅。

在这栋别墅的设计中，马利娜·普蒂洛夫斯卡亚设计工作室实践了他们最爱的自然美的设计理念。蔚蓝的天空映照在泳池的水面上，鸟儿在天空中翱翔，营造出天堂般的美景，还有异域风情的植被……这一切都能令人领略年轻的活力，而别墅室内空间的设计也侧重营造长久幸福的生活。

新古典家居的装饰艺术

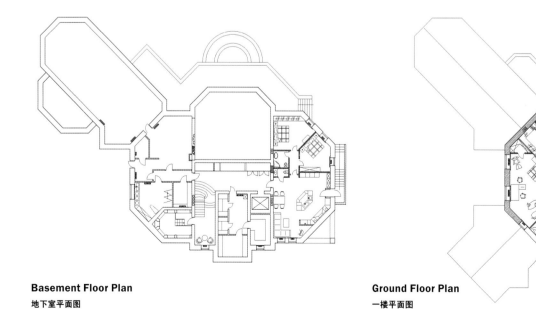

Basement Floor Plan
地下室平面图

Ground Floor Plan
一楼平面图

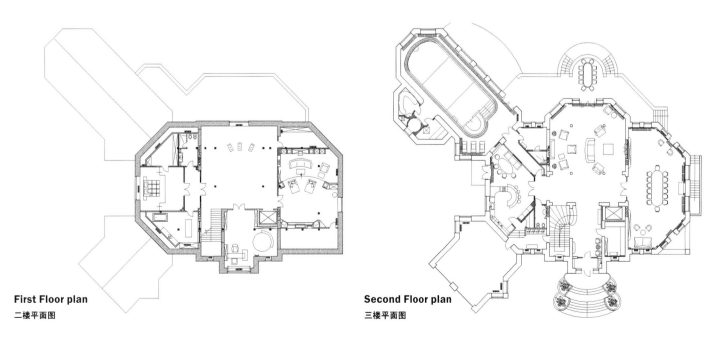

First Floor plan
二楼平面图

Second Floor plan
三楼平面图

新古典家居的装饰艺术

144 DECORATIONS OF NEO-CLASSICAL HOME DESIGN

6. Home cinema with bold and bright colours
7. Swimming pool on the first floor
8. The owner's bedroom
9. Child's bedroom
10. The child's room is painted with patterns with child interest to make the space lively.
11. Bathroom for women
12. Detail of bathroom for women
13. Bathroom for men

6. 家庭影院采用明亮、大胆的色彩
7. 二楼泳池
8. 主卧
9. 儿童房
10. 儿童房墙面上的喷绘让房间更加生动、有趣
11. 女士浴室
12. 女士浴室洗手台
13. 男士浴室

5. Furniture

When the requirement for home design goes higher, furniture has gone beyond its basic function. The relationship between pieces of furniture and the existing environment should always be taken into account. Home spaces grow increasing large, but the use of partition walls is reduced. Instead, furniture is more often used as partitions, playing multiplc roles in home design. In home furnishings design, furniture plays an important part. The choice of furniture is a decisive factor for the establishment of the tone of furnishings design, and other pieces of furnishings should be designed accordingly.

Home layout and environment should be a primary consideration. The shape, size and location of furniture should be decided according to its function in daily use, aiming at an optimal result for circulation

5.家具造型与环境

随着居住环境设计要求的不断提高，家具的作用已经不仅仅是满足使用需要，更要考虑其与存在环境的关系。居住空间越来越大，墙体越来越少，很多隔断也被家具所取代，家具在空间中的作用也越来越大。在家居陈设中，家具也是其中的主体。家具的形式决定了空间的陈设基调，其他陈设品需要围绕这个主题展开。

家具的选择首先要考虑布局和空间的环境。家具的形态、大小、摆放方式都要结合使用要求，使空间的交通、采光、造型都达到最佳的状态，另外也要方便使用并且提高空间的利用率。新古典风格家具相对而言比较精致，造型也比现代家具复杂，所以要有更充足的空间和恰当的环境。客厅中的家具一般以壁炉前的沙发为中心，沙发通常会包括一组多人沙发和一组对称的单人沙发，中间是茶几，沙发上有厚重的软垫，扶手靠背也用织物包裹，椅腿和桌腿都雕刻

and daylighting. Moreover, user convenience and space utilisation should be maximised. Neo-classical furniture is relatively more complicated and delicate than modern furniture, and thus ample spaces and appropriate environment are required. In the living room, furniture design mainly consists of sofas in front of the fireplaces, usually including a group of settee and symmetrical placement of single sofas, with a tea table in the centre. On the sofa we often have heavy cushions, and the armrest and back of sofa are usually upholstered with fabrics. The legs of tables and chairs can be carved into claws of animals, which is particularly common in neo-classical furniture design. Places near the wall in the living room are often placed with side tables or other pieces of furniture, which, besides basic function of storing, can be used to hold furnishings on the top. The pieces of furniture selected should be unique; both delicate neo-classical and Baroque ones will do, but always remember to keep them in accordance with the interior design of the space.

The selection of furniture is also an important element to the establishment of the interior atmosphere. Furniture will have a great influence on the atmosphere, and are also symbols of the characteristics and taste of the whole design, resulting in different psychological perceptions on the user. Neo-classical furniture features dignity and simplicity in shape, often with clear profiles and symmetrical forms, highlighting both vertical and horizontal structures, feeling heavy and strong. As for details, early Baroque and Rococo styles are integrated. For example, you can find gold inlays and curved surfaces, especially on legs, where curves are often adopted. Utility is highlighted, with simple yet elegant forms. Such furniture stands for the requirement of European design on texture and colour in their tradition and history, and at the same time the complicated decoration is much simplified, making the space feel both classical and modern. Of course, the selection of furniture varies according to the style of the interior. Elegant and graceful pieces of furniture from France and simple and lively British ones are both good choices, and there are other styles available to be applied in specific locations. For instance, in the living room for a woman, feminine Rococo furniture can be adopted, but attention should be paid to the holistic style of the room, and the furniture mustn't be overwhelming in the interior.

All the pieces of furniture in the following project are meticulously selected by the designer, provided by famous Italian manufacturers. From the sofas, chairs and the bed, to the wardrobes, bedside tables and the dining table, everything is delicate in detail, and all coexist harmoniously in one space. Seating elements vary greatly, including heavy armchairs in the living room and the settee in the bedroom, all upholstered with luxurious textiles and filled with soft packing. The heavy texture matches up with the dignified style of the design, and the colours and patterns go well with the surrounding soft decoration. There are not many pieces of furniture in the home, but each piece is naturally integrated with the environment.

成动物脚的形状，这是新古典风格常见的家具形式。客厅中靠墙的位置也会摆放边桌或其他类型的家具，这些家具除了有存储作用之外，还会在上面摆放陈设品。家具造型一般会比较有特色，精致的新古典风格或巴洛克风格都很合适，但要与室内的界面设计搭配得当。

家具的选择同时也要为营造空间气氛服务。家具会给空间氛围带来很大影响，也表现了空间的设计特色和品味，让人产生不同的心理感受。新古典风格的家具造型稳重，线条简洁，常用对称的造型，并且突出水平和垂直的结构设计，着重表现结构的力度，在细节处也吸收了前期巴洛克和洛可可风格的造型，有镶金或曲面，尤其是腿部一般会采用曲线，家具本身注重实用性，造型也尽量简洁。这样的家具保留了欧式家具对材质、色彩的要求和丰厚的文化底蕴，又简化了复杂的装饰，使空间既有古典风范又有开放的姿态。当然，家具的选择也可以根据环境有所变化，有法国家具的高贵典雅，也有英国家具的简洁明快，还可以在合适的地方点缀其他风格类型，例如在女性的起居室可以增加柔美的洛可可风格家具，但不能喧宾夺主，影响整体。

本案例中的家具全部由设计师精心挑选，来自于意大利的著名制造商。从沙发、椅子、床，到衣柜、床头柜、餐桌，每一样都有精致的造型，并且能共存在一个空间中。其中椅子的类型比较多，除了客厅中厚重的扶手椅，还有卧室里的小塌，表面全部包裹着华美的织物，并且有柔软的衬垫物，厚重的质感与空间稳重的设计风格一致，在色彩和图案上也有周围的软装饰和谐统一。整个空间中的家具并不多，但都与空间的环境紧密结合在一起。

1. Bedroom in Zvenigorod house
1. 兹维尼哥罗德别墅卧室

Zvenigorod
兹维尼哥罗德别墅

This house is made in classic style, where you can see Marina Putilovskaya's special character. The interior is designed as a harmonic unity of unique details. All the sketches were drawn by Marina Putilovskaya. Each fragment of stucco, each design of a stone or parquet floor mosaic is unique and was made solety for this project. The furniture and the lightning for this house were brought from Italy; the architect found them in classical collections of famous Italian manufactures – Provasi, Jumbo Collection, Beby Group. The floors in the house are made from marble and valuable woods. The restained and tranquil colour palette of the interior is in line with the classical canon. The bedrooms were spacious, with a total area of 800sqm, each with a fireplace and a boudoir. A bathroom is made from stucco work and decoration of mirrors. When discussing with the designers, the clients immediately marked their colour preferences. They asked to make the interior warm and calm. The only exception was made for the blue living room on the second floor.

Main living room with fireplace, blue dining room, owners' bedroom with soft light and velvet fabrics, bathroom, like a casket. Each of these interiors affects all the senses.

The main theme of the house is autumn. The bright golden tones remind the leaves and the sunlight. Very bright spot in this interior is the blue dining room, which resembles the winter.

A perfect effect is made with a mixture of warm colours and mirrors. This effect enlarges the room. The fireplace in the bedroom is less massive than in the living room, but it also has a stucco decoration with floral pattern.

The central element in the living room is a fireplace that suited in the centre of the room. This is a bright spot and a massive symbol of warmth and comfort.

Location Moscow, Russia
Designer Design Bureau of Marina Putilovskaya / Marina Putilovskaya
Photographer Federico Simm
Area 1,400m²

项目地址 俄罗斯，莫斯科
设计师 玛丽娜·普蒂洛夫斯卡亚设计工作室（玛丽娜·普蒂洛夫斯卡亚）
摄影师 费德里科·西姆
项目面积 1,400平方米

这栋别墅采用古典主义风格，在这里你可以看到设计师马利娜·普蒂洛夫斯卡亚独特的设计风格。室内空间既有许多独特的细节，同时又形成一个和谐的整体。所有的手绘图纸都是马利娜·普蒂洛夫斯卡亚亲自绘制的。每一处的灰泥、石材的设计以及镶花地板都是独一无二的，而且是为本案专门制作的。家具和照明设备来自意大利，是设计师从意大利知名制造商的古典系列中挑选的，这些知名品牌包括伯瓦西（Provasi）、珍宝家居（Jumbo Collection）和贝碧集团（Beby Group）等。别墅内的地面采用大理石和珍稀木材。室内低调、宁静的色彩旨在与古典风格相一致。卧室都非常宽敞，总面积达到800平方米，里面有壁炉和化妆室。浴室采用灰泥粉刷，并以镜面装饰。在前期与设计师商讨时，委托客户直接给出了他们偏爱的色彩。他们要求室内的色调要温暖、宁静。唯一的例外是三楼的起居室，设计成蓝色。

2. The floor of the hall is made of marble.
3. The central element in the living room is a fireplace that suited in the centre of the room.
4. Small living room
5. Living room full of natural light
6. The very bright spot in this interior is the blue dining room, which resembles the winter.
7. Dining area
8. Kitchen next to the dining room

2. 大厅地面采用大理石
3. 起居室里的核心元素是位于中央的壁炉
4. 小起居室
5. 起居室阳光明媚
6. 室内空间的亮点是蓝色的饭厅，凸显冬日风情
7. 用餐区
8. 饭厅旁边的厨房

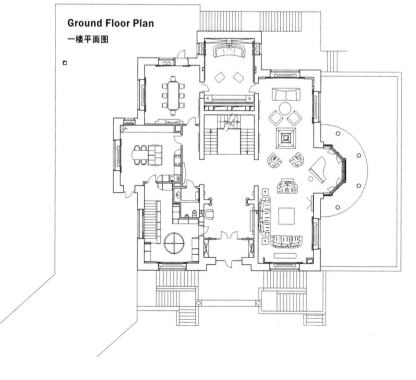

Ground Floor Plan
一楼平面图

First Floor Plan
二楼平面图

Second Floor Plan
三楼平面图

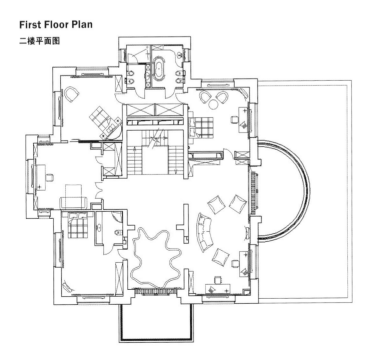

9. Owner's bedroom with soft light and velvet fabrics
10. The tube like a casket
11. Calm swimming pool

9. 主卧有着柔和的光线和温暖的天鹅绒装饰织物
10. 浴缸像个精致的箱子
11. 清新的泳池

主要起居室里有壁炉；饭厅弥漫着蓝色色调；房主的卧室有柔和的光线和天鹅绒织物；浴室小巧玲珑，像个小箱子。这些空间每一个都给人带来强烈的感官体验。

这栋别墅的主题是秋季。闪亮的金色色调是落叶和阳光的颜色。蓝色的饭厅是室内空间的一大亮点，这里代表的是冬季。

暖色调和镜面的结合营造出完美的效果，扩大了空间的视觉面积。卧室里的壁炉没有起居室里那么大，但也采用了灰泥装饰花纹。

起居室里的核心元素是房间中央的大型壁炉，照亮了整个房间，象征了温暖和舒适。

6. Application of Baroque Style in Home Design

The Baroque style has been popular in architecture and interior design in Italy, Austria, Spain and Germany since the 17th century. We can find similar designs in France, Britain and Northern Europe, but slightly different with the previous ones. After the period of great prosperity in the Renaissance, the Baroque style had been developed, and Mannerism came into being, highlighting complicated decorations. Both architecture and interior design began to feature a sculptural beauty. Decorations were mostly inspired by nature, including plants, shells and cartouches. Such decorations can be used everywhere in the interior, especially on the wall and ceiling, with various patterns, sculptural forms and multiple colours. When originally emerged, the Baroque style was mostly applied to religious buildings, featuring fantastic paintings on the wall and ceiling in the interior, making visitors feel as if being in heaven. Baroque furniture is also known for its complicated decoration, sometimes even criticised as too showy. Such furniture is a micro manifestation of the Baroque architecture, featuring dynamic curves, cartouches and columns. In a later period, we can find gildings, coloured paintings and intarsia on the furniture surface. Materials like stone, textile and metal can be used for decoration, making the pieces of furniture already large in size feel more gorgeous and sumptuous.

The Baroque style was originally labelled as a religious trait, luxurious, and full of imagination. Nowadays it is often understood as being equal to complex decorations, and seldom adopted as the style of the whole design in home spaces. However, some of its components and characteristics can be separately used in specific areas, especially the dynamic curves and eye-catching furniture. Curves and ovals are typical forms in the Baroque style, which can be seen in beautiful vaults and complicated treatments of stairs. They help add dynamism to the interior, and also a sense of mystery. Such elements and treatments can be applied in home design, if requirements for the size and ceiling height of the space can be met. Lastly, Baroque furniture is a highly ornamental element, being strongly masculine. In a spacious area, a piece of such furniture can be placed as artwork, completing a vintage style in the space.

In the following project, many Baroque elements are integrated into the British neo-classical home spaces. At the entrance, blue Baroque sofas are placed, echoing with the white Baroque chandelier on the

6.巴洛克造型在住宅中的应用

巴洛克风格是17世纪流行于意大利、奥地利、西班牙、德国等地的建筑和室内设计风格，在法国、英国及北欧等地也有类似的作品，但略有差异。巴洛克风格的设计在文艺复兴鼎盛时期的基础上有所发展，强调手法的繁复装饰，无论是建筑还是室内设计都极具雕塑感。装饰造型一般都取自自然中的元素，花草、贝壳、涡卷饰等，这些装饰遍布在室内各处，尤其是墙面、顶棚，装饰造型不仅繁多并且生动立体，色彩丰富。巴洛克风格在形成之初多用在宗教建筑上，所以，常常会在室内塑造出梦幻般的天棚画或壁画，使人产生天堂般的错觉。室内的家具同样以复杂的装饰著称，甚至有些过于卖弄，这种类型的家具吸收了建筑的特征，有动感的曲线、涡卷装饰、柱式装饰等，到了后期，还有镀金、彩绘、细木镶嵌等技术

ceiling. Chairs in the dining room feature Baroque curves, and the cabinets placed nearby present Baroque mouldings. The furniture and furnishings in the house didn't show all the characteristics of Baroque style; instead, only some qualities are shown in the use of lines and decorations, but traditional Baroque complexity and grandness abandoned. Clear and pure colours and forms are used to slightly adorn the space, making the small house intimate yet full of artistic air. For ordinary houses, the use of Baroque style should be cautious, especially in small to medium sized spaces, where some Baroque traits can be presented in an abstract way to achieve a better ornamental effect.

进行家具表面的加工，石材、织物、金属等都能被用在家具的装饰中，使原本尺寸就很大的家具更加富丽堂皇，气势十足。

巴洛克风格最初有很强的宗教特色，既豪华又富有想象力，而如今，它更多地被用来形容过于繁琐的装饰，在日常的家居中已经很少会整体采用这种风格。但其中的某些特点还是常常会被单独提取出来用在特定的地方，尤其它动感的线条和装饰效果极强的家具。巴洛克风格的造型中经常会用到曲线或椭圆形，例如有着优美弧线的拱顶和复杂的楼梯处理，不仅为室内增添了动态的效果，也带有一些神秘感，在家居设计中同样可以借鉴，只是对住宅的面积和举架高度要求较高。另外，巴洛克的家具有很强的装饰性，并且充满了阳刚的气质，在空间宽敞的地方可以作为一件艺术品点缀，产生复古的效果。

在本案例中，一个有着英式新古典风格的家居空间中融入了多种巴洛克的元素。入口处有蓝色的巴洛克风格沙发，上面装饰着白色巴洛克吊灯；餐厅中的餐椅有巴洛克风格的曲线，旁边的柜子也带有巴洛克的线脚。这些家具或陈设并没有全部继承巴洛克的特点，只是借鉴了巴洛克风格的线条和装饰特色，而抛弃了巴洛克一贯的复杂和宏大，尽量采用简洁纯粹的色彩和造型，在住宅中稍作点缀，使这个空间并不大的住宅既有温馨的气氛，也有很强的艺术效果。对一般的住宅来说，巴洛克风格都不宜过多和过于浓烈，尤其是中小型的住宅，适当地抽象出巴洛克的特点就可以起到很好的装饰效果。

1. Living room in Casa Mataró

1. 马塔罗公寓起居室

Casa Mataró
马塔罗公寓

Behind an old monumental façade Elia Felices creates a cosy home, and at the same time art gallery, located in Mataró, Barcelona.

On the ground floor we find the entrance hall. Materials are mixed: a pair of restored rocking chairs with an upholstered Baroque indigo sofa, and a coffee table of mirror, everything makes an unthinkable mix. The reflections of the white Baroque pendant lamp located in the centre of the space behind uncover old landscape frescoes, slightly framed by a wooden slat lacquered in white, all enveloped by a grey atmosphere. Walls are sober in grey, "le nouveau noir".

From the entrance hall were acceded a guest room, the living/dining room and kitchen.

Two large wooden doors in walnut colour invite you into the guest room, with a 19th-century English style black bed with golden highlights.

There's a sofa with wood trim and a restored old chest of drawers. Raw and white colours confirm the wisdom by neutral tones and create contrast with the grey walls, white roof and beams recoloured in grey.

In the dining room, the dining table with slightly oval glass envelope and central stainless sreel foot creates a contrast with the chairs of Baroque mouldings, satin and floral motifs. Black and white curtains give a scenic nature in the room.

In the kitchen there are a worktop of white stone, central island with a bell suspended in steel and black lacquered furniture.

In the upper hall there are two cubes mirrored as nightstands, two restored Baroque moldings and a sofa with organic mouldings with black velvet, gramophones cushions, chairs, lamp on a Louis XIV style console lacquered in black and a monumental portrait reflected in the great mirror of golden Baroque molding.

The principal bedroom mixes fantasy and serenely classic decor. A majestic bed of walnut with high gloss finish contrasts with the liturgical white of bed linen.

On its opposite side there's a youth twin bedroom, with 19th-century English style black and golden forge with a canopy in white curtains.

Finally, we find the bathroom. Intermixed materials were also used in this room. A large sink in white marble Macael, a dark brown wooden shelf, a silvered reflection mirror and an elegant and sophisticated glass shower screen. The mixture of periods and styles gives personality to the house.

Location Mataro, Spain
Designer Elia Felices interiorismo
Photographer Rafael Vargas
Area 190m²

项目地址 西班牙，马塔罗
设计师 伊利亚·费利斯室内设计公司
摄影师 拉斐尔·瓦格斯
项目面积 190平方米

2. Two large wooden doors in walnut colour invite you into the guest room.
3. The reflections of the white Baroque pendant lamp uncover old landscape frescoes.
4. Mixed materials in the living room
5. The upper hall with two cubes mirrored as tables
6. A monumental portrait makes the hall more classical.

2. 两扇胡桃色大木门连接着客房
3. 房间中央的巴洛克风格白色吊灯在后面古老的风景壁画上形成倒影
4. 起居室采用多种材料相结合
5. 二楼大厅里摆放着方体镜面茶几
6. 肖像画进一步凸显了空间的古典风格

该住宅位于巴塞罗那马塔罗，建筑外观具有古老的标志性特征，伊利亚·费利斯将其打造成一个温馨舒适的家，同时也是一间画廊。

一层设有入口大厅，各种材料混合使用，包括两把经过修复的摇椅，巴洛克风格的靓蓝色软垫沙发，还有一张镜面的咖啡桌，所有这一切构成了一个不可思议的融合。位于房间中央的巴洛克风格的白色吊灯在后面古老风景的壁画上形成倒影，壁画以白色涂漆的木制板条镶框，营造出一种灰白色的氛围。墙壁采用了冷灰色调，是一种"新黑色"。

入口大厅通往客房、起居室和厨房。通往客房的两扇大木门是胡桃色的，带有19世纪英式风格的黑红色，并以金色渲染。屋内摆设着木质装饰的沙发，和经过修复的老式

7. The kitchen with a worktop made of white stone
8. Black and white curtains in the dining room frame the natural scenery outside.

7. 厨房里白石材质的工作台
8. 黑白色的窗帘给餐厅形成一道自然景观

五斗橱。原生色和白色通过中性色调调和，并与灰色的墙面、白屋顶和重新染成灰色的横梁形成对比。餐厅内摆放着椭圆形的玻璃面餐桌，桌脚为不锈钢镜面抛光材质，与巴洛克风格的带有花形图案的缎面餐椅形成鲜明的对比。黑白色的窗帘给房间形成一道自然景观。厨房工作台为白石材质的，挂钟悬挂在钢化材质和黑漆罩面的家具中。

二层大厅中摆放着方体镜面的茶几，两把经过修复的巴洛克风格座椅和一个黑天鹅绒面沙发，上面配有带留声机图案的靠垫，还有一张路易十四时期风格的控制台，黑漆罩面，上面摆放着台灯，墙上挂着具有纪念意义的肖像画，正好反射在对面的巴洛克风格的金色边框的大镜子中。

正对面是一间年轻人的双人间卧室，房间充满了具有19世纪英式风格的黑色和金色，配有白色的窗帘，形成一道天幕。

最后是浴室，这个房间也使用了各种材料混合搭配。大大的白色大理石面的洗手台、深棕色的木制置物架、镀银的反光镜，还有典雅精致的玻璃面淋浴屏风。各个时期和各种风格的混搭形成了这座住宅的个性特征。

Ground Floor plan
一楼平面图

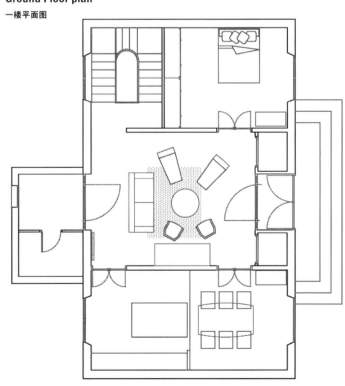

7. French Residence in Black and White

7.黑白之间的法式住宅

The French neo-classicism, originated in the late 18th century, was called the Louis XVI style in the early days. From the late 18th century to the early 19th century, France was the centre of the art of architecture in Europe. Fast social and economical development and the occurrence of the French Revolution all provide the basis for the evolution of neo-classicism. Some archaeological findings offer direct inspirations for neo-classical design.

Different from other regions in the world, neo-classicism in France is extremely delicate, graceful and elegant, with the finest forms and patterns. Firstly, it overturned the showy quality of Baroque and Rococo style, getting rid of their too much decorations and curves, and drawing inspiration from the legacy of ancient Rome and Greece

法国的新古典风格在早期也叫路易十六风格，产生于18世纪末19世纪初。在这一时期，法国是欧洲建筑艺术的中心，社会经济的发展和法国大革命的发生为新古典主义的发展提供了基础，而一些考古发现也为新古典主义提供了更直接的素材。

与其他地区不同，法国的新古典风格更精致考究，强调优雅高贵的气质，造型也更加纤巧、轻盈。首先，它改变了浮华的巴洛克和洛可可风格，取消了过分的装饰和过多的曲线，又从古罗马和古希腊的遗产中吸取了精华，有复古的意味。洛可可风格是形成于法国宫廷中，反映贵族生活的艺术生活，而新古典主义产生于欧洲发生一连串社会巨变的时代，一开始就建立在理性、秩序、创新的基础上，没有完整的规律可循，只有取材于古代艺术。

to present a vintage air. Rococo developed in French palaces as an artistic style reflecting the aristocratic life, while neo-classicism came into being when a series of social upheavals happened in Europe. It was based on rationality, good order, and innovation, without comprehensive principles to follow, but all fragmentally inspired by ancient arts.

Secondly, apart from the rigorous artistic beauty, French neo-classicism also puts emphasis on the utilitarian function. The design of furniture is a good example. Many exquisitely shaped pieces are quite practical in use. Desks popular in that period were designed with mechanical devices to adjust flexibly; boards on two sides can be unfolded as shelf for books; in the middle there are drawers for storage; and the desk top can be adjusted into a sloping surface for easy reading. Besides, such a desk, made of natural wood, is often decorated with bronze components and columns, completing a perfect balance between utility and natural beauty. The dressing table for ladies was also ingeniously designed, with special hidden shelves for the storage of cosmetics. The desk for needlework was designed with double levels. On the upper level, drawers were equipped for the storage of needlework, while on the lower level, when the desk top was unfolded, small spaces were meticulously arranged. All the designs of the furniture reflect wisdom of the designers at that time, and some of them are still in use today.

Lastly, being elegant and graceful, French neo-classical style also features complicated techniques and an open attitude. Apart from the mechanical devices in furniture design mentioned above, some techniques are often used, including carving, inlay, painting and some skills used in architecture. Delicate gilded bronze, carved festoon and rose ornaments… all can be integrated to present the noble elegance of French neo-classicism. Moreover, in these complex decorations and techniques, some Oriental elements can be incorporated, including landscape paintings, coloured drawings and lacquering techniques from China, demonstrating the open attitude of French neo-classicism.

The following project is a simple yet artistic French neo-classical residence. Black and white were used as the two main colours, creating an elegant and exquisite space. Compared with the romance and delicacy of conventional French neo-classicism, the design of this residence is much simplified, without much decoration on surfaces, or gold or silver treatments. Instead, a rigid layout of space, rational proportions and moderate floral and fret patterns are used to showcase a typical French neo-classical style. Light colours and simple arrangement of space respond with the surrounding continuous hills, demonstrating a perfect co-existence of home space and the natural environment.

其次，法国新古典风格除了严谨的艺术美之外，同时也关注实用功能，例如家具的设计，许多造型精巧的家具都有很具体的实用功能，在新古典时期流行的写字台带有机械装置，可以随意调整状态，翻开两侧台面可以用作搁板放书，中间有抽屉储物，桌面还可以调节成斜面方便阅读。并且，写字台有铜制的装饰和柱式点缀，用实木制作，简单淳朴之中有自然之美，并且与实用功能完美结合。另外还有供女士们使用的梳妆桌，有专门的暗格放置盥洗用品；用于放针线的针线桌，有上下两层，不但有抽屉可以放针线，下面的桌面还可以翻开，有若干小空间。这些家具都体现了当时设计者的智慧，有些类型的家具一直沿用至今。

另外，法国新古典风格在典雅优美的基础上，也有复杂的工艺和开放的态度。除了上面所说的家具的一些机械装置，在室内装饰中还常常会用到雕刻、镶嵌、绘画以及建筑上的一些工艺，精美的青铜镀金，雕刻的垂花饰、玫瑰花饰等融为一体，呈现出了法国新古典风格高贵优雅的态度。在这些复杂的装饰和技艺中，新古典风格还吸收了来自东方的装饰元素和特殊的工艺，如中国的山水画、中国彩绘、漆工艺等，这也从一个侧面反映了它开放包容的精神。

本案例是一个简洁又艺术的法国新古典风格住宅，设计师通过简单的黑白两种主色创造出了一个淡雅精致的空间。相对于法国新古典风格的精致浪漫，这间住宅更加简化，没有过多的界面装饰，没有镶金描银，只是保留了法式风格严谨的布局，合理的比例和点到即止的花卉装饰、回纹饰。淡雅的色彩和简洁的布局与住宅周围连绵的丘陵映衬在一起，充分展现了环境与家居的共生。

1. Kitchen in Twomey Cuntry House
2. Media room detail, Twomey Country House

1. 图米乡间别墅厨房
2. 图米乡间别墅多媒体室家具陈设

Twomey Country House
图米乡间别墅

The clients had seen a previous project the designers had completed in Woollahra and fell in love with the unique balance between the traditional and edgy which was created largely through the use of black and white. However, being located in the beautiful Southern Highlands of NSW the clients wanted a softer version of this that would highlight the gorgeous location and the uniqueness of looking out across rolling hills, a direct contrast to their previous Sydney abode. The owners were envisaging a French flavour to the house as well as furnishings and interiors that were slightly weathered and which could withstand the rough and tumble of a young family without sacrificing any of the drama and interest that they loved about the Woollahra house.

The original architecture of the circa 1990's faux federation home was dark and did nothing to make the most of the views and light. In response to the clients' brief the designers set about creating a home that has a sophisticated patina which will age gracefully with the family. Architecturally one of the most significant changes was to open up the house to the view across the hills through the addition of French doors across the whole living room, kitchen and master bedroom spaces. The other major change was to create a larger arch opening between the kitchen and dining rooms which gives the whole space a more cohesive, open and light atmosphere.

For the designers, the interiors introduced a major departure from previous projects by placing most of the pattern and decoration onto the floors whilst keeping the walls and furnishings clean and simple. This was achieved by using a highly geometric black and white carpet throughout the corridors and bedrooms and then geometric rugs in the entry, living and dining rooms over timber floors. In the conservatory this use of patterned flooring was continued by using stunning handmade Moroccan hexagonal tiles.

Location Sutton Forest, New South Wales, Australia
Designer Greg Natale Design
Photographer Anson Smart
Area 800m²

项目地址 澳大利亚，新南威尔士州，萨顿森林镇
设计师 格雷格·纳塔莱设计工作室
摄影师 安森·斯玛特
项目面积 800平方米

本案的委托客户曾经见过格雷格·纳塔莱设计工作室在沃拉拉区打造的一所别墅，他们非常喜爱其黑白两色搭配呈现出的传统与现代交融的风格。然而，由于本案的地点位于新南威尔士州的南部高地，所以他们想要风格上更柔和一些，凸显地段的优势和屋外独特的风景——起伏的群山，这跟他们之前在悉尼的住处大不相同。房主希望这座别墅能够呈现出一丝法式风情，室内装饰和家具陈设都略微呈现出历经风霜的感觉，要能够适应一个年轻家庭日常生活的杂乱无章，同时又不失他们所喜爱的沃拉拉区别墅的那种风格。

这栋别墅始建于20世纪90年代，室内比较昏暗，之前无法利用室外的景色和光线。为了满足委托客户的要求，设计师打造的室内空间散发着年深日久积成的光泽，并将随着这户人家一起经受时间的考验。建筑结构上最主要的改造就是打通别墅毗邻山脉的一侧，利用山景的视野。设计师采用了法式落地双扇玻璃门，贯穿起居室、厨房、主卧等空间。另外一项较大的改造是在厨房和饭厅之间增加了一个大型拱形开窗，让整个空间更加连贯、宽敞、明亮。

对设计师来说，本案的室内设计跟他们之前的设计的主要不同之处在于：主要的图案和装饰都用在地面上，而墙面和室内陈设则相对干净简洁。走廊和各个卧室全部采用几何构图的黑白地毯，入口门厅、起居室和饭厅的地板上也铺设了几何图案的小块地毯。日光室里也延续了地面图案的使用，采用手工制作的摩洛哥六边形地砖。

3. Master bedroom
4. Entrance foyer
5. Entrance hall
6. Entrance hall with geometric rugs
7. Media room
8. Kitchen and dining area
9. Kitchen detail
10. Kitchen joinery detail
11. Clean and simple kitchen
12. A large arch opening between the kitchen and dining room

3. 主卧
4. 入口门厅
5. 入口大厅
6. 入口大厅采用几何图案的地毯
7. 多媒体室
8. 厨房和饭厅
9. 厨房特写
10. 厨房细木工艺特写
11. 厨房简约大方
12. 厨房和饭厅之间设计了大型拱形开窗

13. French flavour lounge
14. Lounge full of light
15. Sofa and lamp in the lounge
16. Lounge detail
17. Comfortable sofas in the library
18. Reading area in the library and study
19. Well-organised library and study

13. 法式风情的休闲区
14. 休闲区洒满阳光
15. 休闲区沙发和灯饰特写
16. 休闲区一角
17. 图书室里设置了舒适的沙发
18. 图书室和书房里的阅览区
19. 图书室和书房整洁有序

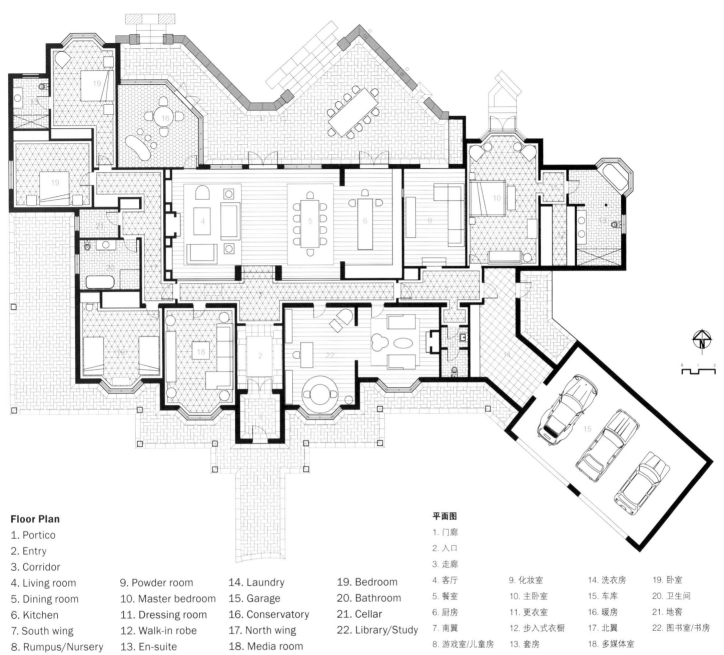

Floor Plan

1. Portico
2. Entry
3. Corridor
4. Living room
5. Dining room
6. Kitchen
7. South wing
8. Rumpus/Nursery
9. Powder room
10. Master bedroom
11. Dressing room
12. Walk-in robe
13. En-suite
14. Laundry
15. Garage
16. Conservatory
17. North wing
18. Media room
19. Bedroom
20. Bathroom
21. Cellar
22. Library/Study

平面图

1. 门廊
2. 入口
3. 走廊
4. 客厅
5. 餐室
6. 厨房
7. 南翼
8. 游戏室/儿童房
9. 化妆室
10. 主卧室
11. 更衣室
12. 步入式衣橱
13. 套房
14. 洗衣房
15. 车库
16. 暖房
17. 北翼
18. 多媒体室
19. 卧室
20. 卫生间
21. 地窖
22. 图书室/书房

20. Bedroom for guests
21. Bathroom in master en-suite
22. Bath in master en-suite
23. Spacious master bedroom
24. Dressing room in master en-suite
25. Detail of master bedroom
26. Walk-in robe in master en-suite

20. 客房
21. 主卧套房里的浴室
22. 主卧套房里的浴缸
23. 宽敞的主卧
24. 主套房化妆间
25. 主卧床头陈设特写
26. 主套房步入式衣橱

An artistic space must answer to requirements for a healthy living environment.

Apart from function and decoration, a comfortable living space is closely related to the physical environment. Usability of a living space is determined by rationality of the layout, which, more importantly, will affect the health of family members.

A space comes to our eyes first, and light is important for its visual effect. In a home, natural light can improve the quality of spaces, while artificial light can be used to create various atmospheres. Good ventilation is necessary for a healthy living environment and for the reduction of energy consumption. Appropriate interior greening can purify indoor air and adjust temperature; it is also helpful in organising spaces and guiding directions. Besides

艺术的环境即是健康的环境

舒适家居空间的打造除了要有完善的功能和艺术的装饰之外，与物理环境也有很大的关系。环境的合理程度影响着家居空间的功能使用，更重要的是，它关系着家居环境是否科学，关系着家庭成员的健康。

从最基本的视觉开始，光环境影响着家居环境的每一个角落，自然光的融入是提升环境质量的基础，人造光则是打造家居气氛的重要手段；而通风则是实现卫生健康的家居环境和降低能耗的必要条件；适当的室内绿化能净化空气，调节气温，也能起到空间的组织和引导作用；而除了室内本身的建设，将室外的景观引入室内，实现内外的交流更能体现设计的多样性；实现了这些物质要求之外，人的心理需求同样不能忽视，环境心理学在理论上为打造适合人心理需求的空间提供了基础，也为设计师在实际操作中提供了方向。

CHAPTER 4
NEO-CLASSICAL ART FOR A BETTER LIVING ENVIRONMENT
新古典家居的环境艺术

indoor spaces, the outdoor environment is also a part of home design. Views of the outdoor landscape can be introduced inside to establish a connection between the inside and the outside. Besides these physical requirements, there are also psychological needs. Environmental psychology is an emerging overlapping domain developed in recent years, which focuses on the relationship between essential environment and people's behaviour. It provides references for architects and designers to create psychologically healthy environments.

An ideal living environment must be one that is healthy, and for a healthy environment, communication with nature is essential. Such kind of communication is invisible, but can be realised through visible designs. Creating pure and natural home spaces is the goal of such communication, and it is also a kind of art.

一个理想的家居环境首先要是健康的环境，健康的环境离不开与自然的交流，这种交流是无形的，但却可以通过有形的设计实现，让家居生活更自然更纯粹是交流的目的，也是一种艺术。

1. Natural Light in Every Room

Light is the soul of a space. Without it, nothing in the space can be perceived. Natural light changes greatly in a year and even in one day. Therefore, the use of natural light can be complicated. However, the influence of natural light, especially the dramatic play of light and shadow, is an indispensable element for a good home design.

Spaces with natural light are particularly important for homes. Firstly, different spaces in a home require different lighting conditions. Spaces with distinct functions or accommodating different people need various lighting designs. So in the design process, space function and people's preference for lighting conditions should be taken into account. Generally speaking, natural light, more or less, is needed in every space in a home, but the demand for intensity of

1. 自然光线进入每一个房间

光是一个空间的灵魂，没有光，空间中的一切形态都不会被人感知。自然光在一年四季，一天之中都有不同的变化，所以，利用起来也更复杂，但自然光的作用和光影形成的变化却是空间中不能缺少的元素。

家居环境中，自然光环境更是一个重要的部分。首先，家居环境应该满足不同空间对光的要求。不同功能的空间和不同的人对自然光的要求不同，所以，设计中应该充分考虑空间的性质和居住在空间中的人的喜好。一般来说，家居环境中每个空间都需要有一定的自然光，但需要有不同的强度。新古典风格形成的年代住宅窗户都比较小，采光差，室内环境比较昏暗，如今的住宅都尽量采用全明的朝向，尤其是一些别墅，常常会加大窗户的面积，或采用整体的玻璃幕墙，增大采光的强度。在客厅、餐厅、厨房这些地方尽量保

lighting varies. Many neo-classical houses were built in early days and have relatively small windows which allow for less natural light in, resulting in a poor indoor lighting quality and dim interior spaces. In contemporary residences, the direction is set to make best of natural light. Especially in some new houses, windows would be enlarged or, even a glass curtain wall would be used for an entire elevation, to intensify natural lighting in interiors. Living rooms, dining rooms and kitchens should be as bright as possible, especially in neo-classical homes, which usually enjoy a large floor area, the living room often faces a huge window to avoid feeling gloomy due to the dusk of the room. Sufficient natural light is also needed in the bedroom, where ample natural lighting is good for disinfection. This is particularly important for bedrooms for the elderly and children, which should be placed to enjoy good daylighting. However, sunlight allowed into the bedroom shouldn't be too strong, so the openings needn't be too large. Besides curtains, louvres or blinds could be adopted for sun shading.

Moreover, the optical properties of interior finishes decide the final effect of colours, forms and ornaments. In neo-classical houses, various materials would be used in the interior. There are usually heavy fabrics and textures, sometimes with gilded decorations. Therefore, sufficient natural light is definitely necessary. The size, location and shape of openings should be considered carefully to ensure the best visual effects of all elements in the interior. Direct light should be controlled and adjusted in particular to avoid glare or overheating. Floor-to-ceiling windows and arch or circular openings are often used in neo-classical style houses. Extensive glazing can bring in abundant natural light, but heavy curtains are needed sometimes to provide shade or keep out the cold. Arch openings bring out the typical classical ambiance, and curtains are also often needed to match up with the elegant shapes.

The project presented here is a castle that occupies a total floor area of more than 2,000 square metres. Though working with a large area and many rooms, the architects managed to produce a layout that allows natural light into each room. The living room has an open layout on both sides, enjoying ample sunlight at all times. In the dining room and the kitchen, rectangular windows are adopted, small but bright and open, producing perfect daylighting effects. Traditional arch windows can be found in the study on all sides. In response to the somehow dazzling light coming in, large areas of panels are used, with calm and dignified colour and texture. Arch windows also appear in bedrooms, smaller in size, with curtains that have the matching curving lines for both flexible shading and a soft atmosphere. With appropriate area of openings and suitable interior textures, the architects took full advantage of natural light and avoided its disadvantages.

持明亮，尤其是对新古典风格来说，面积一般较大，客厅一般会正对大窗，避免昏暗环境使人产生烦闷的情绪。卧室同样需要充足的阳光，阳光能保证卧室中良好的杀菌效果，尤其是老人或儿童的卧室，更需要布置在有自然光的朝向。但卧室的阳光不宜太强烈，不需要太大的采光口，窗口除了窗帘外也可以增加百叶等遮光挡光材料。

另外，室内的色彩、造型、各种装饰效果都需要通过表面材料的光学特性表现出来，对新古典风格来说，室内的各种材料很多，有时还会有镀金等装饰，织物也比较厚重，所以更应该有充足的光源，对采光口的大小、位置、形式也该谨慎考虑，保证室内各种元素的表面效果达到最佳，尤其要对直射光进行必要的控制和调节，以免引起室内眩光或过热。并且，新古典风格的住宅常常会采用大型的落地窗或拱形窗、圆形窗。大面积落地窗采光效果好，但也要搭配厚重的窗帘，在必要的时候遮光御寒，而拱形窗有浓厚的古典气质，同样需要搭配能突出其轮廓的窗帘让古典的气质更完美。

本案例是一个超过2000平方米的城堡住宅，虽然面积很大，房间很多，但设计者却通过精巧的构思使每个房间都有自然光的照射。外面的客厅是两面通透的格局，随时有充足的光线。厨房和餐厅采用了采光效果最好的长方形窗，面积不大，但明亮通透。书房是传统的拱形窗，并且环绕了整个空间，但室内的装饰采用了大面积的镶板，色彩和材质都以稳重为主，能使光线在室内显得不过于刺眼。卧室也有小面积的拱形窗，搭配了同样具有弧线的窗帘，遮挡了部分光线，也使卧室的环境变得更加柔和。恰到好处的采光面积和合理的室内材质让这个家居环境充分利用了自然光的优势，规避了它的缺点。

1. Exterior of 40 Beverly Park

1. 贝弗利公园40号住宅外景

40 Beverly Park
贝弗利公园40号住宅

Located on a two-acre parcel of land in a Los Angeles gated community, this estate takes advantage of modern building technologies while holding true to the design characteristics that make a chateaux of France's Loire valley so intriguing.

With careful consideration to the home's orientation, the design allows for abundant natural light to enter every space and enables the concepts of passive solar and thermal mass to work well. Natural ventilation is used in collaboration with thickened walls selectively clad in natural limestone to help alleviate rising energy costs. Similarly, the location of the pool is also situated to take advantage of the prevailing summer breezes and provide cooling to the outdoor covered loggia spaces.

Conceived from a 600-year-old French chateaux, the environmentally conscious techniques that take advantage of the sun, wind currents and building orientation transport this residence into the 21st century.

Inside, the concept of bringing new ideas to an old style remains. Unlike the original chateaux of the Loire valley, the rooms flow from one to another, organised along wide galleries and thus allow for an understanding of the spatial relationships within the design.

Each project brings with it its own challenges. By combining new concepts from both a design and environmental standpoint, a home set in its deeply rooted historical precedents has been propelled into the present day.

Location Los Angeles, USA
Designer Landry Design Group
Photographer Erhard Pfeiffer
Area 2,322m²

项目地址 美国，洛杉矶
设计师 德瑞设计团队
摄影师 埃哈德·菲佛
项目面积 2,322平方米

2. Stairs and ceiling
3. Living room
4. Kitchen

2. 楼梯与天花
3. 起居室
4. 厨房

该住宅位于洛杉矶一处封闭式社区中，占地两英亩，设计师利用现代建筑技术和独具匠心的设计，将其打造成为法国卢瓦尔河谷迷人的城堡。

充分考虑其家用性能，该设计利用大量的自然光，使自然光覆盖到每个区域，并且实现了被动式太阳能和热质量的理念。自然通风设施与天然石灰岩覆面的增厚墙板相结合，缓解了日益增加的能源消耗。同样，泳池所在位置也利用了夏日的盛行风，为户外凉亭空间带来丝丝凉意。

设计构思来源于具有600年历史的法国城堡，通过各种环保技术，利用了太阳能、风能和建筑朝向将其打造成具有21世纪特征的住宅。室内空间的设计将新的理念融入到古老风格之中。与原来的卢瓦尔河谷城堡不同，各个房间沿着一条宽敞的走廊排列，设计中的空间关系非常明晰。

每个项目都有其独特的挑战性，该项目的挑战之处在于通过结合新的设计与环保理念，将这座深具历史特征的住宅打造成更加适合现代社会居住的住宅。

5. Round hall
6. Study upstairs
7. Wine tasting room
8. Bedroom with fireplace
9. Master's bathroom
10. Luxury bathroom
11. Swimming pool

5. 圆厅
6. 楼上书房
7. 品酒室
8. 带壁炉的卧室
9. 主浴室
10. 奢华浴室
11. 泳池

新古典家居的环境艺术 | 187

Sketch
手绘图

2. Combined Lighting

Artificial lighting, as a significant part of home design, is getting more and more attention. Artificial light is used not only to fulfill the basic function of lighting up a space, nor as a complement to natural light; rather, it helps complete a creative combination with brightness and darkness, light and shadow, in a way of art. Meanwhile, lighting designs with different intensities, styles and effects would have certain influences on people living in the space, both physiologically and psychologically. Therefore, the rationality of lighting design is particularly important in home spaces.

According to functions and characters of spaces such as the living room, dining room, bedroom and bathroom, different rooms in a house have their distinct requirements for lighting conditions. In a living room you might spend your leisure time, have entertainments, or receive a visitor; thus a complicated light system is required and the selection of lamps should be careful. Especially in neo-classical style homes, some valuable ornaments and furniture need special lighting to achieve a good visual effect. Crystal chandeliers are always preferred in neo-classical designs, but for large living rooms, extra spot lighting would be needed. For example, you could put a floor lamp beside the sofa, or a lamp on the table. Lights from floor lamps can be reflected on the ceiling and then well distributed to fill the whole room, producing a softer effect. Table lamps are set for particular spots, whose shapes should go along with the style of the living room. Apart from these, wall lamps are also preferred in neo-classical design. In a living room they are often adopted on the peripheral walls or on the background wall of the TV set. Wall lamps not only provide spot lighting, but also help in establishing an intimate atmosphere in the living room.

A relaxing atmosphere is needed in the dining room, and with suitable lighting, attention would be focused on the dining table. Usually a main chandelier would be placed above the dining table, but the light source shouldn't be too strong for the eye and direct light on man should be avoided. Soft, yellow light can produce a graceful luster for the table, food and tableware. If connected with the dining room, the kitchen should have a brighter lighting environment to produce more enthusiasm, and spotlights on the ceiling are recommended.

Bedrooms are spaces for sleep and rest, so the lighting environment should be warm and intimate. The main light doesn't need to be

2. 家居中的复合照明

作为家居设计中的一个重要部分，人工照明越来越被人们所重视，灯光已经不仅仅是出于照明的需要，也不仅仅是自然光的补充，而是室内明与暗、光与影的组合，是一种艺术形式。同时，灯光的不同强度、方式、效果，都会对人产生一定的生理和心理影响，所以对家居空间来说，尤其需要合理的照明设计。

根据功能的不同和空间的性质，客厅、餐厅、卧室、浴室等不同位置对照明的设计有不同的要求。客厅有休闲、娱乐、会客等多种功能，所以照明系统也比较复杂，灯具的选择也很重要。尤其是对新古典风格来说，一些珍贵的装饰品和家具等需要相应的灯光才能有好的效果。水晶灯是一般新古典风格的首选，但对大面积的客厅来

very bright, but its form or shape should be elegant and intricate to go with the neo-classical style of the room. Lamps on bedside tables would be used rather frequently, and vintage ones are recommended. If the owner of the house often reads on bed, it would be better if the lighting intensity and angles of the table lamp could be adjusted. In neo-classical style interiors, sometimes symmetrical wall lamps would be used on both sides of the bedhead, as a complement for lighting as well as for decoration. In bathrooms, functionality is the primary concern. For an ordinary bathroom, waterproofing ceiling lamps are mostly preferred. If neo-classical style is desired, droplights with intricate or interesting shapes could be used, complemented with symmetrical lamps beside the mirror. Sometimes ceiling lamps can be replaced by wall lamps placed on various spots.

The project selected here features an eclectic combination of classicism and modernism and diversity in lighting design. There are both luxurious crystal chandeliers and simple table lamps, wall lamps, spotlights and droplights. In the living room, a large crystal chandelier plays a central role, while wall lamps are used to complete a warm and elegant atmosphere with soft yellow tones. Gorgeous crystal chandeliers are also adopted in the bedrooms, combined with contemporary table lamps on bedsides, achieving a balance between classicism and modernism. The foyer is one of the most unique spaces in the home, where numerous spotlights are used to highlight the huge stained glasswork on the ceiling. The whole space is flowing with light and colour, revealing the artistic taste of the master.

说，还需要一些局部照明，例如沙发边上的落地灯或桌上的台灯。落地灯的光线可以反射到棚顶再均匀地洒到每个角落，使光线更柔和。而台灯是针对部分角落的光源，在造型上也需要配合客厅的风格。另外，局部的壁灯也是新古典风格常用的形式，客厅中壁灯一般会安排在周围的墙面上或是电视背景墙上，壁灯能起到局部照明的作用，也能增添客厅亲切和谐的气氛。

餐厅需要轻松愉快的气氛，照明设计也会尽量使人把注意力集中到餐桌上。一般会在餐桌上方设计主灯，但光源不应过于集中和耀眼，防止灯光直接照在人身上。一般柔和的黄色光能使餐桌、食品、餐具等色泽更柔美。如果跟厨房相连，厨房需要更明亮，更能使人有积极性的照明环境，可以在天棚上安排局部的射灯增加照明效果。

卧室是睡眠休息的场所，照明环境以宁静温馨为主。主灯不需要太过明亮，但在造型上可以尽量美观精致以配合新古典风格。两侧床头柜上的床头灯使用频率会比较高，可以选择造型复古的台灯，如果有读书的习惯，还应该带有调整亮度和角度的功能。对新古典风格来说，有时还会在床头位置增加对称的壁灯作为补充的光源，也可以作为装饰。而对浴室来说，实用性是最重要的。对普通浴室来说，防水的吸顶灯是最好的选择。如果需要塑造新古典风格，可以选择造型简洁别致的吊灯，再加上对称的镜前灯。也可以用各部分的壁灯代替顶灯。

本案例中的设计结合了古典和现代的双重风格，在照明上也很有多样性。有豪华的水晶灯，也有简洁的台灯、壁灯、射灯、吊灯。客厅采用了大型的水晶灯和壁灯相结合，柔和的黄色使客厅的气氛温暖优雅。卧室同样是造型华美的水晶灯，配合了现代造型的床头台灯，实现了古典和现代的结合。而最有特点的大厅采用了许多小射灯，烘托了天花板上的大型玻璃彩绘，使整个大厅流光溢彩，也表达了主人的艺术品位。

1. Hall in Godolevski apartment

1. 高德莱温斯基公寓门厅

Godolevski
高德莱温斯基公寓

It's located on the Parkway ring – one of the most romantic places of old Moscow. The apartment is located in an ancient house, constructed in Modern Art style.

The smart hall serves as an interior axis – at the left is located the drawing room and the kitchen, and on the right is the bedroom, study and a bathroom of the owners. In the hall the ceiling attracts attention with its stained-glass window with characteristic Morden Art style ornament- its clouds and flowers. The floor also is in Morden Art style. Specially for the hall's floor architect Maria composed the ornament similar to stylised fish scales. It was created at the Italian factory according to the art drawings of the designer. The architect chose marble together with the hostess of the apartment. The hostess possesses a delicate art taste and so the interior is created by a courageous colour palette. Shades of a cowberry, a lavender, and a lilac are combined with green.

The study-library is executed in classical style with the carved polished tree of noble breeds. The drawing room and kitchen-dining room are connected by a wide aperture. In the drawing room on both sides of a mirror are two hand-painted panels. Natural shades are combined with upholsteries and curtains and all this creates the pleasant atmosphere in the spirit of the Moscow aristocratic salon of the beginning of the 20th century. All furniture is modern, but classical in style.

Location Moscow, Russia
Designer Baharev&Partners
Photographer Zinur Razzutdinov
Area 1,280m²

项目地址 俄罗斯，莫斯科
设计师 巴哈莱夫联合设计公司
摄影师 金努尔·拉祖迪诺夫
项目面积 1,280平方米

2. Floor of the hall with ornament similar to stylised fish scales
3. In the drawing room on both sides of a mirror are two hand-painted panels.
4. Detail of the drawing room
5. Detail of the wall in the drawing room
6. The drawing room and kitchen-dining room are connected by a wide aperture.

2. 门厅地面有鱼鳞状的装饰图案
3. 客厅中镜子的两侧镶嵌着两块带有手绘图案的木板
4. 客厅一瞥
5. 客厅墙面特写
6. 客厅与厨房餐厅相连，中间以宽大的门拱相隔

7. Kitchen and dining room
8. Classical style study-library
9. The study-library is executed with the carved polished tree of noble breeds.
10. Courageous colour palette in the bedroom
11. Owner's bathroom
12. Guest's bathroom

7. 厨房和餐厅
8. 古典风格的书房−图书室
9. 书房−图书室选用了精雕细琢的高品质抛光木材
10. 卧室大胆的色彩运用
11. 主卫生间
12. 客用卫生间

Floor Plan
平面图

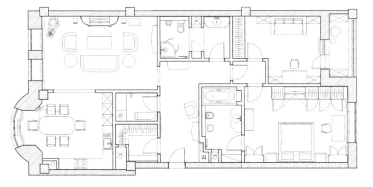

该公寓位于百汇环——莫斯科从前最浪漫的区域之一。该公寓位于一座具有现代艺术风格的古老建筑之中。

精小的门厅作为室内的轴线，左侧为客厅和厨房，右侧为主人卧室、书房和浴室。门厅中天花板上的染色玻璃窗非常引人注目，上面印有云和花形图案，是具有现代艺术风格的装饰。地板也具有现代艺术风格。门厅的地板也很特别，上面的装饰图案是玛利亚建筑师仿照鱼鳞的形状设计，由意大利的工厂根据设计师的图纸生产制造的。建筑师与女主人一起为这所公寓选择了大理石制品。女主人拥有一种清淡典雅的艺术品味，室内设计中色彩的大胆搭配也归功于她，越橘、薰衣草和紫丁香的颜色与绿色相结合使用。

书房-图书室是以古典风格设计的，选用了精雕细琢的高品质抛光木材。客厅与厨房餐厅相连，中间以宽大的门拱相隔。客厅中镜子的两侧镶嵌着两块带有手绘图案的木板。天然的窗纱与窗幔和窗帘相结合，为这个20世纪初期的具有莫斯科贵族气派的沙龙渲染一种轻松愉快的氛围。所有的家具都是具有古典风格的现代家具。

3. Ventilation

Today when residences suffer from high density and industrialisation, ventilation becomes important for the quality of our living spaces, and ventilation solutions are getting more and more attention from both architects and their clients. Ventilation affects our living environments mainly in two ways. Firstly, the residence is relatively a closed space, where the indoor air could be polluted by the furniture and furnishings, metabolic processes of the inhabitants, burning of fuels, etc. These pollutants would cause serious problems to our health if not treated properly, and ventilation can act as the most effective solution. Good ventilation provides fresh air for the indoor environment and more oxygen for the inhabitants, and at the same time helps clear pollutants in the interior and reduce the density of indoor microorganism. Secondly, ventilation is a way to keep the

3. 家居环境的通风

在住宅越来越高密度、工业化的今天，通风情况直接影响着家居环境的品质，如何解决住宅的通风问题也越来越受到人们的关注。通风对家居环境的影响主要表现在两个方面：首先，居室是一个相对封闭的环境，室内的家具、装饰用品；人体的新陈代谢；燃料的燃烧等都会造成室内的空气污染，如果不加以处理会严重影响人们的健康。而通风是解决这种室内污染的最有效办法，有效的通风提供了新鲜的空气，也保证了人体所需的供氧量，同时也排除了室内的污染物，降低室内的微生物密度。其次，通风也是维持室内舒适环境的一种手段。在气温高的夏季，可以通过通风减少室内的热量，增加人体与空气间的对流散热和汗水蒸发散热，使人体在室内能感受到相对合适的温度。

通风的方式通常有自然通风和机械通风两种方式。自然通风主要依

interior space comfortable. Especially in hot summers, ventilation can do a really good job of reducing indoor heat, promoting heat loss through circulation of the air and through evaporation of sweat. In such a space the temperature could be adjusted to a relatively appropriate and comfortable level for human.

Ventilation can be classified into two categories: natural ventilation and mechanical ventilation. Natural ventilation mainly relies on the use of doors and windows, which are usually set in the phase of architectural design, but there are still ways to enhance the effect of natural ventilation in interior design. In neo-classical style designs, large arch doors and windows are often adopted, usually casement windows for better ventilation and cooling effects. If ventilation conditions need to be enhanced, some walls can be adapted to allow for air circulation and wind flow. In some houses, sometimes large viewing windows would be added, which allow for maximal ventilation effect with effective indoor and outdoor air circulation, without disturbing architectural aesthetics of the house. All in all, doors and windows with various forms and opening modes should be adopted to ensure natural ventilation in all kinds of weather.

In some spaces in a residence, such as the kitchen and bathroom, extra mechanical ventilation is necessary. In all spaces where ventilation needs to be enhanced due to outdoor environments or layouts of the interior, mechanical ventilation would be helpful. It is less dependent on conditions of seasons or climates, and can work steadily and regularly as a complement to natural ventilation. It should be noted that the equipment and wiring required for mechanical ventilation ought to be considered when the overall layout is planned.

The following project is a single house, in which rectangular and arch windows as well as arch doors are adopted to achieve a good ventilation effect. The sizes of the windows are maximised, and thus each room is brightly lit with natural light, especially the public space for family gatherings, where a number of doors and windows are used for circulation of air between inside and outside. The high ceiling further enhanced the flowing of air. In public spaces where many people come together, good ventilation would keep the environment comfortable, and in a residence it would be good for the health of family members.

赖门窗，门窗的大小、位置等一般在建筑设计阶段就已经确定，但在室内设计中还是可以通过一定方式增强自然通风的效果。新古典风格通常采用大面积的拱形门窗，并且常用推开窗设计，有利于通风和散热。如果想加强通风效果，还可以通过改造可塑性强的墙体，使房屋格局中形成空气对流通道，形成曲线的对流风。一些别墅有时会采用大面积的观景窗，可以在不破坏外观的情况下，保留通风的最大值，让室内外空气有效流通。总之，应尽量选择多种形式和多开启方式门窗，这样无论什么样的外部气候条件都能达到开窗通风的目的。

在家居空间的一些特殊环境（如厨房、卫生间等），还需要额外的机械通风，其他地方如果因为室外环境或房型等问题也可以通过机械通风的方式增强通风效果。机械通风方式受季节和气候因素影响较小，并且稳定均匀，可以作为自然通风的补充。但在整体规划时就要考虑机械通风的设备和布线等问题。

本案例是一座独栋住宅，为了达到良好的通风条件，设置了方形窗、拱形窗和拱形门，窗户的面积也达到了最大的值，室内的每个房间都宽敞明亮，尤其是用于家庭聚会的公共空间，多个门窗形成了室内外空气的对流，让空气有流进也有流出，较高的举架也使空气的流动更通畅，对聚集人群较多的公共空间来说，良好的通风条件不仅使环境更舒适，也最大程度保证了家庭成员的健康。

1. Exterior of Round Hill Road Residence
1. 朗德山路住宅外景

Round Hill Road Residence
朗德山路住宅

This Georgian stone home set on 1.94 hectares in the exclusive Round Hill section of Greenwich marked the designer's fourth collaboration with this client. The designer's goal for the design was to warm and enliven the very classical, formal space to accommodate a vibrant family of seven.

The family had spent the past decade in a transitory state, having lived in Singapore, London and the New York area, and they were eager to settle into a home that felt personal and inviting. The eight-bedroom home offered exquisite features – sawn oak floors, French doors with leaded glass, high ceilings, a lovely two-bedroom pool house – but needed a thoughtful design strategy to give its abundant spaces a cohesive flow.

To make the large-scale rooms less imposing, the designer began by layering textured wall coverings, beautiful custom millwork and a unifying palette of rich taupe and slate blue, a client directive. An immediate point of reference was the iconic Blue Bar in London's Berkeley Hotel. Like this residence, the Berkeley seduces you with traditional charm that belies the chic, sophisticated space within. The designer loves how the Blue Bar's classical architectural elements are modernised with high gloss paint.

To create a dynamic contrast, the designer made the public spaces bright and airy, while family gathering spaces were given a clubby, lived-in patina. To illustrate, the Family Room began as a rather grand space (historically known as a Great Room) – the goal was to infuse it with warmth and texture while enabling it to multi-task as family room, entertainment centre, children's study and transitional hub. First, the designer dressed the arched ceiling in board and batten, as well as cross-beam millwork, to smooth the transition to the walls. Along the back wall, the designer installed a walnut built-in that invokes the richness of old library cabinets and provides easily accessible storage. The walls were papered with loden green grass cloth and complementary trim, adding plaid wool draperies on the French doors along with a wool sisal-like weave carpet, scaled to reveal the beautiful wood flooring along the perimeter.

The decor reflects the family's global influences, with cool contrasts throughout – note the graphic black and white photography set against their collection of traditional French and English antiques.

Location Greenwich, Connecticut, USA　项目地址 美国，康涅狄格州，格林威治
Designer S.B. Long Interiors　设计师 S.B.朗室内设计工作室
Photographer Don Freeman　摄影师 唐·弗里曼
Area 21,200m²　项目面积 21,200平方米

2. Entrance hall
3. Modernised kitchen
4. Breakfast room
5. In the dining room an immediate point of reference is the Blue Bar in London's Berkeley Hotel.

2. 入口门厅
3. 现代化的厨房
4. 早餐室
5. 饭厅的设计借鉴了伦敦伯克利酒店里著名的蓝调酒吧

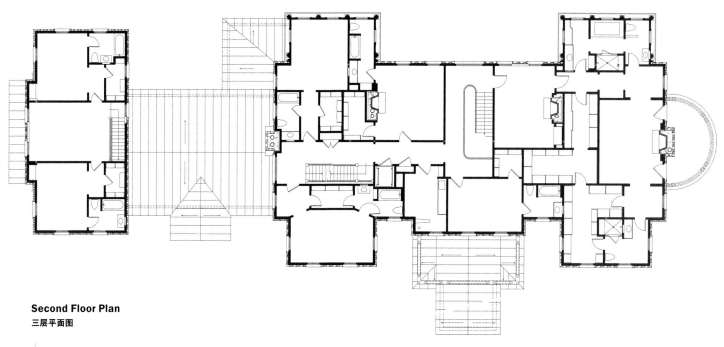

Second Floor Plan
三层平面图

NEO-CLASSICAL ART FOR A BETTER LIVING ENVIRONMENT

这座英国乔治王朝时代风格的石屋，占地约1.94公顷，位于格林威治朗德山高档住宅区。这是本案设计师第四次与这位委托客户合作了。设计师的目标是让这栋住宅充满古典风格的室内空间更显温暖、更具活力，满足7口之家的使用需求。

这户人家过去10年中不断在搬迁，先后曾在新加坡、伦敦和纽约等地居住，所以他们特别渴望能够安定下来，有一个属于他们自己的温暖的家。这栋住宅有8间卧室，原来的装修已经很精致了，包括橡木地板、法式落地双扇玻璃门（安装的是铅条玻璃）、高高的天花板，等等。泳池边还有一座小屋，里面有两间卧室。不过，这栋住宅还需要按照统一的设计策略来让各处空间的联系更加紧密和流畅。

这栋住宅空间很宽敞，但是过大的体量显得不够人性化。为此，设计师在墙面上增加了充满质感的材料，还有定制的精美的木工制品，主色调采用灰褐色和石蓝色，这是委托客户指定的颜色。设计师借鉴了伦敦伯克利酒店里著名的蓝调酒吧（Blue Bar）。跟这栋住宅一样，伯克利酒店也是外观上采用了古典风格，而室内则是时尚、现代的装修。设计师特别喜爱蓝调酒吧古典建筑元素与高光泽度涂料相结合的处理方式。

室内公共空间宽敞明亮，而家庭活动空间则私密、舒适，两者形成鲜明的对比。尤其是家庭活动室，装饰非常豪华，设计目标是营造充满质感的温暖空间，使其融入住宅的整体风格，同时赋予该空间多种功能，包括家庭活动室、娱乐中心、儿童学习室等。首先，设计师采用板条来装饰拱形天花板，增加了木质横梁，使天花自然地过渡到墙面。房间后面的墙壁安装了胡桃木的固定壁橱，营造出老式图书馆书柜的厚重感，存放物品也很方便。墙面采用深橄榄绿色的麻布装饰，搭配精致的镶边，法式落地双扇玻璃门采用格子花呢作为装饰织物，此外还有毛料的纺织地毯，看起来像西沙尔麻一样，地毯四周露出精致的木质地板。

这栋住宅的装修体现出这户人家的环球居住经历，室内处处呈现出鲜明的对照，比如黑白摄影作品和家里收藏的法国和英国古董放在一起。

6. The family room is clubby.
7. Sitting room
8. Warm and enlivened living room
9. His office with black and white photography
10. Elegant her office
11. Master's bedroom
12. Little girl's bedroom

6. 家庭活动室俨然一间俱乐部
7. 起居室
8. 客厅
9. 男主人办公室采用黑白摄影照片来装饰
10. 女主人办公室精致典雅
11. 主卧
12. 小女孩卧室

4. Indoor Greening

Green plants increasingly play indispensable roles in interior design. With plants in a living space, we will feel close to nature, with both visual and psychological enjoyments. In addition, plants act as an air purifier and temperature regulator, thus becoming one of the most efficient ways to improve indoor environment quality. In a residence, plants can be used in different ways according to differences on location and requirement, and various kinds of plants can be combined flexibly. The overall structure, as well as consistency of colours and forms should be considered in the process of indoor greening design.

Planting, just like any other element in interior design, should be used with proper proportions and rational layouts, especially in

4. 家居绿化设计

绿色植物在室内设计中有着特殊的作用，它能满足人们接近自然、返璞归真的需要，给人带来视觉上的享受和心理上的调节，还能净化空气、调节室内的温度，是改善室内环境最有效的手段。在家居空间中，每个位置都可以根据不同的目的和作用采取不同的绿化方式，家居中的绿化就是把各种植物组合起来，使植物的基本特征在居室内得到最好的发挥，并且在环境中体现出构图的合理、色彩的协调和形式的和谐。

在室内的设计中，绿化装饰与其他元素一样同样需要构图的合理，也就是布置的均衡和比例的适当。尤其是在对比例有严格要求的新古典风格中，为了使室内的所有元素显得规则整齐，庄重和谐，必须使空间内所有局部元素的设计服从于整体的设计理念。对植物来说，形态、大小、位置都会影响到它们在室内产生的视觉效果，为

neo-classical style, which has a strict requirement for proportions. In order to create a harmonious, solemn interior, all the elements should be consistent with the overall atmosphere. As for plants, the visual effects vary greatly due to different locations and sizes. Such factors should be taken into account for the creation of a visually pleasing environment. For instance, if plants with different sizes are placed on the two sides of the fireplace, a visual unbalance would be caused, disturbing the symmetrical aesthetics of neo-classicism. Plants with similar forms and sizes are preferred to create a neo-classical living room. In narrow corridors with low ceilings, tall plants shouldn't be placed; otherwise the space would feel crowded. Instead, small plants in pots or vases are recommended.

Functionality is another important factor for indoor greening. Some plants have their destined locations and functions due to their inherent characters. For example, in China particular plants are perceived with fixed connotations: cyclamens are a symbol of distinguished guests, and therefore are often placed in living rooms, while asparagus ferns are related to literature and thus suitable for the study. Apart from cultural preferences, plants should be located according to lighting and humidity conditions of the interior. Tall and lush plants would produce shades in the interior, so they are often placed in corners or beside pieces of furniture. In bathrooms where natural lighting is usually poor and indoor air humid, rustic plants would be a good choice.

Flowers and vines are preferred in neo-classical design as important decorative elements to complement the elegant, graceful atmosphere. For indoor greening in neo-classical home design, frequently used are tall trees and romantic vines; bright and colourful flowers are also preferred such as beautiful tulips and fragrant roses.

The selected project is a single house situated in a well-landscaped environment with beautiful views. Lush trees and plants create comfortable leisure spaces around the house as well as in its small courtyard. The green ambiance continues inside the house. In the neo-classical interior spaces, you can find plants everywhere. In the living room two plants with the same height as the mantel are placed on both sides of the fireplace, keeping to the principle of symmetry like the columns in the room. There are vines in the kitchen, potted plants and flowers with various colours in the dining room and bathroom, and tall plants as decoration in the corners near the staircase. A vigorous living environment is thus created, echoing the beautiful scenery outside while being in accordance to the taste of neo-classicism.

了保持在构图上的稳定和舒适，就要在视觉上使这些植物保持平衡和协调。例如，在客厅壁炉两边，如果摆放大小相差悬殊的植物，就会产生视觉上的落差，破坏新古典风格对称的原则，而形态和大小都相近的植物更能表达出新古典风格客厅和谐庄重的气氛。再如，在狭窄并且天棚较低的走廊位置摆放高大的植物会产生一定的压迫和拥挤的感觉，而如果换成小型盆栽或花瓶里的插花就会更精致、合理。

绿化设计的另一个原则是功能的要求。一些植物本身的特性决定了它们的功能和位置，例如，一些有美好寓意的植物可以放在相应的位置，像客厅的仙客来、书房的文竹等。同时，植物的摆放也要考虑室内的光线、湿度。如果是高大并且枝叶茂密的植物放在室内会容易产生阴影，所以一般都会放在角落或家具旁边，而卫生间这样光线不足并且比较潮湿的地方则需要一些生命力较强的植物。

新古典风格本身就会运用一些花朵、蔓藤等元素做装饰，再加上其高雅的气质要求，在室内的绿化中，常用的植物通常是高大的乔木、浪漫的藤蔓植物，还有郁金香、玫瑰等色彩鲜艳的花朵。

本案例是一栋本身就处在一个优美环境中的独立建筑，并且还带有一个小庭院。住宅外和庭园中郁郁葱葱的植物创造出了一个舒适的休闲空间，这种优美和舒适在室内也有所体现。室内的设计虽然是典雅的新古典风格，但却随处可见植物的存在。客厅的壁炉两边是两株与壁炉架等高的植物，与客厅中的柱子一样，严格地遵循对称的原则。厨房的橱柜上方有蔓藤植物，餐厅、卫生间有色彩各异的盆栽和鲜花，楼梯角落也有高大的植物点缀，这些形态各异的植物营造了一个生机盎然的家居环境，与室外的优美景色相映成趣，同时，在设计上也迎合了新古典风格的趣味。

1. Front elevation of Summerland

1. 夏日乐园正面外景

Summerland
夏日乐园

This Santa Barbara-styled, Spanish-Mediterranean-inspired home masterfully showcases classic proportions and understated elegance. The stunning barrel roof tile accents the Spanish-Med architecture. Clean minimal use of moldings allows the solid wood doors and textural stone risers to be the focal point of the entry.

When entering the front door, the two-storey foyer creates drama at first glance. The open floor plan is masterfully designed to define and separate spaces using columns and arched entries. The floor plan focuses on allowing access and views to the courtyard and its garden from all of the main level's living areas.

All functional rooms are located on the main level and three en-suite bedrooms are located upstairs. The family room is designed with a unique curved glass wall to soften views through the home by leading the eye to the beauty of the garden. The living room opens to the outside on both sides allowing for nice breezes and expanded entertainment space. The gourmet kitchen and butler's pantry allow for ease in entertaining in the adjoining dining room. The lower level master bedroom is spacious. The master bathroom has his and her bath areas that share a shower.

Eclectic use and combination of furniture allows the home to have a collected appearance. Custom draperies are designed to emphasise architectural style and detailing.

The lanai, pool, and firepit provide a comfortable and beautiful space for casual entertaining. The space also allows the reflection of the elegance of the home's rear elevation.

Location Winter Park, Florida, USA
Designer Phil Kean Designs
Photographer Harvey Smith Photography
Area 5,173m²

项目地址 美国，佛罗里达州，温特帕克
设计师 菲尔·基恩设计工作室
摄影师 哈维·史密斯摄影工作室
项目面积 5,173平方米

这栋别墅呈现出美国加利福尼亚州西南海岸的圣巴巴拉风格，设计灵感来自西班牙/地中海风情，完美展现了古典主义的比例和低调的奢华。尖屋顶的造型凸显了西班牙/地中海建筑的特征。线脚的使用简洁、干净，突出了坚固的木门和整齐的石头台阶，使木门和台阶成为入口的焦点。

从正门进入别墅后，双层举架高度的门厅立刻让人眼前一亮。开放式的空间布局，利用柱子和拱形入口，不着痕迹地界定出若干空间。平面布局的设计侧重室内与庭院的连接，一楼的所有生活空间都与庭院相连，从室内各处也都能看到庭院。庭院里有个小花园，景致优美。

所有的功能性空间都设置在一楼，三间套房式卧室设置在楼上。家庭活动室的设计以弧形玻璃墙面为特色，将视线引至美丽的庭院花园，让室内空间的视觉效果更加柔和。起居室两边都直接与户外相连，温和的清风能够吹进室内，也扩大了娱乐空间。烹饪厨房和管家备餐室里，各种设备一应俱全，足以满足在旁边的饭厅招待宾客的需求。中间层的主卧非常宽敞，男女主人各有自己的沐浴空间，共享一个淋浴间。

别墅内采用了多种多样的家具，兼收并蓄的风格让室内呈现出多样化的面貌。量身定做的装饰织物凸显了别墅的建筑风格和精致的细节处理。

门廊、泳池和篝火炉营造出美观舒适的户外休闲空间。从这里也可以欣赏这栋别墅后方的优美景致。

2. Garden in the courtyard
3. Courtyard with swimming pool
4. When entering the front door, the two-storey foyer creates drama at first glance.
5. Living room on the ground floor
6. The living room opens to the outside on both sides.
7. Breakfast nook and family room
8. Butler's pantry
9. Dining room

2. 庭院花园
3. 庭院里有泳池
4. 从正门进入别墅后，双层举架高度的门厅立刻让人眼前一亮
5. 一楼起居室
6. 起居室两边都直接与户外相连
7. 早餐室和家庭活动室
8. 管家备餐室
9. 饭厅

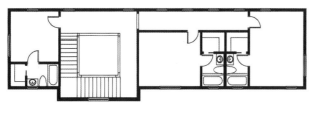

Upper Level Floor Plan
上层平面图

10. Gourmet kitchen
11. The kitchen allows for ease in entertaining in the adjoining dining room.
12. Master bedroom is spacious.
13. The master bathroom has his and her bath areas that share a shower.

10. 烹饪厨房
11. 厨房里各种设备一应俱全，足以满足在旁边的饭厅招待宾客的需求
12. 宽敞的主卧
13. 主浴室里，男女主人各有自己的沐浴空间，共享一个淋浴间

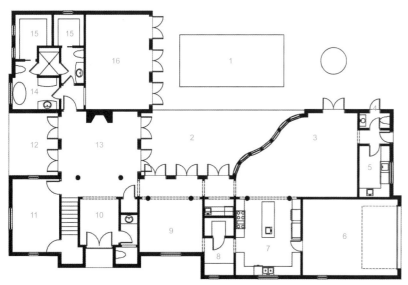

Main Level Floor Plan 主层平面图

1. Pool	9. Dining room	1. 泳池	9. 餐室
2. Lanai	10. Entry	2. 门廊	10. 入口
3. Family room	11. Office	3. 家庭活动室	11. 办公室
4. Powder room	12. Porch	4. 化妆室	12. 走廊
5. Laundry	13. Living room	5. 洗衣房	13. 客厅
6. Garage	14. Main bathroom	6. 车库	14. 主浴室
7. Kitchen	15. W.C.	7. 厨房	15. 卫生间
8. Butler's room	16. Master bedroom	8. 管家房	16. 主卧室

5. Connection between Interior and Exterior

No longer how spacious or gorgeous are our homes, we are still longing for nature, particularly today when many people live in highrises in dense cities, connection with the exterior becomes more and more needed. In such circumstances, establishment of connections between the interior and exterior is necessary in home design.

Such connections are often realised through the use of transitional spaces or viewing windows. Transitional spaces in a residence include the terrace, balcony and foyer, which are very important for our daily living experience. Researches have shown that people are inclined to stay in such semi-open spaces because here they can enjoy both privacy and views towards the outside. However,

5. 室内与室外的交流

不管家居环境多么宽敞和美轮美奂，人们依然保留着对自然的亲近和热情，尤其是如今很多人生活在城市的高层住宅中，跟外界的交流变得越来越少，在家居的设计中体现出室内外的联系就显得尤为重要。

家居中室内和室外的交流一般会通过过渡空间或观景窗来实现。家居中与室外有关的过渡空间包括露台、阳台、玄关等位置，在家居环境中，这些位置对于人的生活体验来说很重要，有研究表明，人们更愿意在这种半开敞的空间逗留，因为在这样的空间中，人们既可以参与或观察到外面的各种活动，又有一定的私密性。但在设计中，这些位置却很容易忽略，尤其是别墅这样过渡空间复杂的类型，而正是这部分空间在很大程度上起到了室内和室外的交流作用。过渡空间把室外的元素（例如阳光、水、植物等）引入到室

such spaces are often ignored in interior design, especially in houses where transitional spaces can be extremely complicated, but it is these spaces that act as connections between the inside and the outside. Transitional spaces bring in exterior elements such as sunshine, water and plants, and make people feel outside even if they are being inside their houses. In home spaces where most rooms are enclosed, they are the best locations to establish a natural, relaxing atmosphere. In heavily decorated neo-classical homes, such an atmosphere is particularly needed. In the design of transitional spaces, textures and colours should be carefully used to achieve a natural transition. For example, tones similar with the outdoor environment, surfaces with grains, and natural materials such as vines can be used to highlight the link with nature. The interior layout can be designed with interconnected spaces to create a multiple layering structure, in which the exterior can be a part.

Viewing windows are also an important tool to establish the connection, bringing more light in and enlarging our views. For apartments in highrises or residences that lack balconies or terraces, viewing windows can serve as the most effective way to expand the vision. In ancient China, such windows were already used to "borrow" outside sceneries in, as they would say in Chinese, to bring people close to nature visually. In modern architectural design, we choose the location and size of such windows according to the surrounding context, and in this way the outside scenery can be brought in as a framed painting. Floor-to-ceiling windows and bay windows can do a good job.

The project here is a highrise residence, lacking in transitional spaces such as terraces, and thus lost an opportunity to be connected with the exterior. Fortunately, large viewing windows are adopted to compensate. In the living room and dining room, windows as high as 4.5 metres provide sufficient natural light, and bring in extensive sceneries. In this way, the otherwise dignified and solemn living room heavily decorated in neo-classical style, becomes light, open and transparent. The gorgeous, traditional interior contrasts with the busy streets outside. In such a living room, guests can drink cocktail while appreciating the exciting world out of the window. Glasses blur the border between inside and out and take lavish cityscapes as a setting for your home.

内，让人即使在室内也能感受到外界的气息，对一成不变的家居空间而言，是放松情绪、感受自然气息的最佳位置，对装饰浓重的新古典风格来说，尤其需要这种轻松的氛围调节。在过渡空间的设计中，可以通过空间的材质、色彩强化室内外的衔接，例如用与室外环境类似的色调、带有纹理的材质、藤麻等天然的材料展现与自然的联系，另外，还可以通过空间的相互穿插、渗透增强空间的层次感，并且使室内和室外相互连通、贯穿，呈现出丰富的层次变化。

观景窗也是实现室内与室外交流的重要手段。观景窗可以增加室内的自然光线，更能扩大人的视野，对于那些高层住宅或缺少阳台的住宅来说，是在室内最重要的观景方式。在中国古代就有了这种"借景引入"的手法将户外的景观引入到室内中，使人在视觉上亲近自然。在如今的建筑设计中，根据周边自然环境设计观景窗的位置和大小，可以把周围的风景引入室内，作为室内自然的风景画，大型的落地窗、飘窗等都可以起到这样的效果。

本案例是一处高层住宅，缺少像露台这样与室外交流的过渡空间，所以，大型的观景窗就格外重要。在客厅和餐厅的位置，高达4.5米的窗户提供了充足的自然光，也引入了大面积的外景，使客厅一改新古典风格的沉重，显得格外通透。室内华丽的新古典风格映衬着室外繁华的城市景观，在客厅这样的地方，客人们可以一边喝着鸡尾酒，一边欣赏窗外缤纷的世界，室内和室外通过有形却没有障碍的玻璃窗实现了交流，室外丰富的景观变成了家居中最好的布景。

1. Dining room in Sokol

1. 索科尔公寓饭厅

Sokol
索科尔公寓

This penthouse was designed for a talented, bright and just wonderful woman. Design Bureau of Marina Putilovskaya wanted to bring the light in their work, creating a beautiful interior, "a palace for the queen". 4.5-metre-high windows did exactly what they wanted – the maximum light and softness. As befits a palace, the interior has a lot of elements, including the marble, crystal, mirrors, moldings, and stained glass windows.

This penthouse is a perfect place for public and popular person, and that is why such a plan was made for this flat. So the owner will not have any problems if she invites her friends for a cocktail. The ground floor takes over the function of public area with magnificent views. One of them, at the entrance, offers the most beautiful piece of flat – entrance area with a crystal staircase and a combined living room area. The first floor is occupied by private areas including a bedroom, a spacious walk-in closet and private bathroom. The bathroom area faces the window, with an alternative to focus – built-in ceiling plane of the TV.

All these were made by Marina Putilovskaya sketches. In the interior unique materials were used such as marble, which is used in furniture and floor, combined with artistic murals, onyx, which is used in owner's and guest toilets, author molding, forging, and exclusive doors. Bright, mirror image of the house formed colours. The main colour is tender silver with elements of gold. A feeling was made that the interior is floating in the air.

Location Moscow, Russia
Designer Design Bureau of Marina Putilovskaya
Photographer Design Bureau of Marina Putilovskaya
Area 195m²

项目地址 俄罗斯，莫斯科
设计师 玛丽娜·普蒂洛夫斯卡亚设计工作室
摄影师 费德里科·西姆
项目面积 195平方米

2. Detail of stairs
3. Entrance area with a crystal staircase
4. Kitchen decorated with marble
5. Dining room

2. 入口处的楼梯有着水晶般的台阶
3. 楼梯是入口处的亮点
4. 厨房采用大理石装饰
5. 饭厅

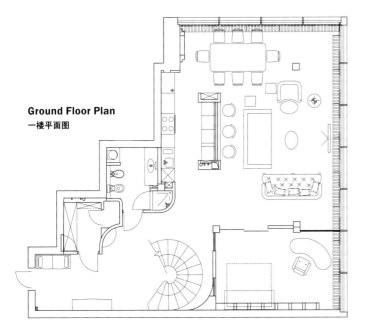

Ground Floor Plan
一楼平面图

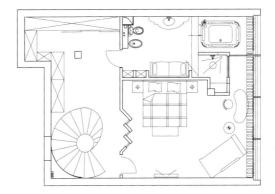

First Floor Plan
二楼平面图

新古典家居的环境艺术 | 225

这是一间顶层公寓，房主是一位聪明又美丽的女性。马利娜·普蒂洛夫斯卡亚设计工作室希望用他们的设计为她打造出"女王的宫殿"一般的室内空间。高达4.5米的开窗实现了这一点，带来充足的光线和温暖的氛围。为了营造宫殿的感觉，室内采用了许多设计元素，包括大理石、水晶、镜面、线脚和彩色玻璃窗等。

这间顶层公寓对于热爱交际的人来说非常理想，这也是为什么设计师为之设计了这样的空间布局。所以，如果房主邀请朋友们在家开派对的话不会有任何问题。一楼主要是公共空间，拥有良好的视野。入口处是这间公寓里最美的空间之一，连接着客厅，楼梯有着水晶般的台阶。二楼主要是私人空间，包括一间卧室、一间宽敞的步入式衣橱以及私人浴室。浴室里有一扇落地窗，还安装了电视，从天花板上悬垂下来。

所有的设计都是由马利娜·普蒂洛夫斯卡亚亲自绘图。室内采用了独特的材料，比如家具和地面采用了大理石，搭配艺术壁画、玛瑙装饰品（用在主卫和客卫）、原创的装饰嵌线、锻造品以及独特的大门。建筑明亮的色调决定了色彩的选择。主色调采用柔和的银色，搭配金色元素，营造出一种室内空间飘浮于空中的感觉。

6. 4.5-metre-high windows give the living room maximum light and softness.
7. Master bedroom on the first floor
8. Public space on the first floor
9. The owner's bathroom area faces the window.
10. Unique materials were used in the guest's bathroom such as artistic murals and onyx.

6. 高达4.5米的开窗给客厅带来充足的光线和温暖的氛围
7. 二楼主卧
8. 二楼楼梯口
9. 主浴室里有开阔的落地窗
10. 客用卫生间里采用了独特的材料，如艺术壁画和玛瑙装饰品等

6. Environmental Psychology and Home Design

The quality of our living environment is determined not only by the comfort or aesthetics provided through interior design; rather, psychological experiences of living in the space can be very important. Environmental psychology is about researches and theories in this field. It deals with how to make environments fulfill our needs from a psychological point of view, aiming at improving our living quality through satisfying our psychological requirements.

The environment exerts a subtle influence on our moods, and different environments lead to different moods. In a playful children's room with bright colours, you would feel relaxed as if going back to childhood. In a solemn study you can easily concentrate and contemplate. That is to say, in the design of every space, its function,

6. 环境心理学与家居设计

家居环境质量的优劣除了与设计的舒适美观有关系之外，人对环境的使用心理也是一个重要的领域，而环境心理学是支撑这个领域的理论基础。环境心理学从人的心理角度探讨了什么样的环境是符合人需求的环境，通过探讨人的心理需求来改善居住环境。

环境会对人的情绪产生潜移默化的影响，不同的环境也会让人产生不同的情绪。色彩缤纷、充满童趣的儿童房会让人心情轻松，犹如回到童年；肃穆的书房会让人精神专注，冷静思考。所以，在每个空间的设计中，都应该考虑其用途、使用者和对使用者带来的心理影响，以使用者在其中的心理状态为出发点。

在一个空间中，首先对人造成心理影响的是色彩的运用。色彩不仅能产生视觉上的艺术享受，也能使人产生联想和感情，直接影响环

1

users and psychological influences on the users should be taken into account, taking the mental state of the user as the starting point.

Firstly, a space first exerts psychological influences on us through colours. The use of colours creates not only artistic visual enjoyment, but imagination and sentiment, setting the basic tone of the atmosphere. Warm colours bring out intimacy, bright colours result in passion and dynamism, and calm colours such as black, white and grey can create a soothing ambiance. The psychological influences of these colours are not decisive, but they can affect our mental state unconsciously.

Secondly, different materials have their effects on our perception with different textures, forms, and patterns, establishing particular atmospheres which set a basic setting for our emotional state. Metals with shining lustre have a typically modern air, but they tend to feel cold; woods with natural grains would provide a warm and intimate ambiance; soft wools help create comfortable spaces; while natural materials such as cotton, linen and rattan remind us of the coexistence with nature. With suitable selection and combination of materials, we can produce any desired atmosphere that leads to different emotions.

Thirdly, light and shadow play a significant role in producing certain psychological effects. Human beings have an intrinsic inclination for light, and we are inclined to move from the dark area to the bright. Light has multiple functions besides lighting up a space. It can affect our perception of other elements and set various tones for the interior. In home spaces, we can create different lighting environments, be it bright, soft, or dynamic, in which we will feel open, intimate, or passionate. Appropriate lighting environments can enrich the otherwise monotonous living space, and greatly change our perceptions towards the space.

The selected project is a house with beautiful outdoor sceneries. Though situated in an urban context, the house is designed with a new style that combines the metropolitan location and rustic feeling. In the interior, luxurious and intricate neo-classical elements are used, balanced with an air of rusticity produced by careful selection and combination of materials, forms and colours. A neutral colour palette goes through the whole interior, which not only produces a soft and elegant ambiance, but reminds us of the tranquil country life. Besides, abundant natural stone and wood are used for decoration, and forms of the petal and leaf are adopted for furniture and lamps. Plants are placed here and there. All in all, the designers successfully created the atmosphere just desired, in which the scene of countryside constantly comes to our mind.

境的气氛。暖色能使人产生亲切温暖的感觉，明亮热烈的颜色能让人觉得热情、活跃，而黑白灰这样的色彩则更加安静、柔和。这些色彩的影响并不是决定性的，但却在不知不觉中对人的心理产生着影响。

其次，不同的材料也能强化人的心理。室内各种材料的不同质感、构造、纹理能渲染出不同的气氛，同时影响人的心理。色泽明亮的金属材料有现代感，但也会让人产生冰冷的感觉；有自然纹理的木材温馨古朴，更有人情味；柔软的羊毛制品能让人觉得温暖舒适；而天然的棉、麻、藤则更容易让人感觉到自然。通过材料的选择和组合，可以有效地烘托空间的气氛，塑造出符合人情绪的环境。

另外，光影在环境中也会对人造成心理影响。人类固有的趋光心理使人们在空间中倾向于从暗处向明处流动。但是光的功能也是多元的，除了照明之外，它还能创造出不同的环境气氛，也能赋予室内其他要素不同的内涵。在家居环境中，光环境有的明亮，有的柔和，有的动感。在不同的光环境下，人们会感觉到或是开阔明朗，或是温馨浪漫，或是富有激情，可以说，光环境的设计能改变环境固有的单调，也改变着人对环境的理解。

本案例中的住宅室外有非常美丽的景观，虽然处在城市中，但设计师却想探索出一种在城市中有乡村感觉的新风格。因此，在室内的设计中不仅有新古典的精致华丽，也通过材质、色彩、形态的选择让人感觉到了淳朴的自然气息。整个空间都采用了统一的中性色调，这种色彩不仅能使人感受到柔和淡雅，也能让人联想到宁静的乡村生活。另外，设计师还运用了大量的天然石材、木材做装饰，在形态上也尽量采用花瓣、树叶等造型的家具、灯具，再加上各个角落里的天然植物，设计师用室内的各种元素暗示了他想表达的宗旨，也让人很容易地想到了乡村的景象。

1. Courtyard in Hashemian Family Residence

1. 哈什米安别墅户外庭院

Hashemian Family Residence
哈什米安别墅

This little jewel box of a home is the designer's personal family residence. The house was built in the 1960s as a semi-custom/tract house. The latest complete renovation on the house was completed at the end of 2010. Basically the house was taken down to the bare structure, and built back up again.

To set the background, Palos Verdes is a small city tucked in an isolated hill not too far away from Los Angeles. The city is situated on a hill, with many winding streets and a very unique European/Mediterranean charm. With this project the designer took a chance and explored a new different style; he really wanted to take advantage of the outdoor views and wanted to preserve that countryside feel of the city. Therefore, the house has a very calm monochromatic feel with punches of contrast in colour and texture.

There is nothing boring about this house and people who have visited the house time and time again have always said that they notice something new each time they visit.

In contrast to working with larger scale homes with a lot of livable square footage, in smaller homes like this residence with a total area of about 335 square metres, one has to become creative and use every square foot to the optimum. There is no room to be wasted and each area has to serve its function to the fullest degree. In this home the designer has created a soothing atmosphere in the private areas. One can feel at ease unwinding in the master bedroom at the end of a long work day as it is a sanctuary fulfilling the mind and the soul.

Location Palos Verdes Estates, California, USA
Designer Casa Siena
Photographer Michael Garland Photography
Area 335m²

项目地址 美国，加利福尼亚州，派洛斯福德庄园
设计师 锡耶纳家居设计公司
摄影师 迈克尔·加兰摄影工作室
项目面积 335平方米

这间小别墅是设计师自己的家庭住宅，始建于20世纪60年代，是当地成批建造的住房之一，属于半定制的住宅。最新一次彻底翻修于2010年底完工。这次翻修将房屋拆得只剩基本的框架结构，相当于彻底重建。

加州的派洛斯福德庄园是一座小城，位于一座孤立的山脉之上，离洛杉矶不远。这座城市位于山上，所以市内有许多蜿蜒的街道，有着欧洲或者地中海地区特有的风情。在本案的设计中，设计师因地制宜，探索了一种新的设计风格，充分利用户外风景，意在保留这座城市的乡村风情。因此，这栋别墅具有一种静谧的乡村氛围，在色彩和质地上形成鲜明的对照。但这栋别墅丝毫不会令人觉得乏味，到过这里多次的人总是会说，他们每次来都会有新的发现。

像这样的小型别墅，总面积只有约335平方米，其设计与大体量住宅有很大不同，后者有大量生活空间可供设计发挥，而小别墅的设计必须注重创意，利用好每一平方米的空间。没有浪费的余地，每个空间必须充分发挥其功能。在这栋别墅中，设计师在私密的空间内营造了一种舒缓的氛围，在一天漫长的工作结束后，可以回到主卧，享受身心的放松。

Floor Plan
1. Garage
2. Study/Den
3. Kitchen
4. Dining room
5. Family room
6. Bar
7. Deck
8. Living room
9. Bedroom
10. Master bedroom

平面图
1. 车库
2. 书房/私室
3. 厨房
4. 餐室
5. 家庭活动室
6. 吧台
7. 平台
8. 客厅
9. 卧室
10. 主卧室

2. Entry hallway
3. Family room
4. Countryside feel family room
5. Living room
6. Stone fireplace in the living room
7. Dining room
8. Natural stone is used in the kitchen.
9. Kitchen detail
10. Work area in the kitchen

2. 入口门厅
3. 家庭活动室
4. 家庭活动室呈现出浓郁的乡村风情
5. 客厅
6. 客厅里的石砌壁炉
7. 饭厅
8. 厨房采用天然石材
9. 厨房一角
10. 厨房工作台

11. Master bedroom
12. Guest's bedroom
13. Detail of the master bedroom
14. Sitting area in the master bedroom
15. Bathroom vanity
16. Stone wall in the bathroom
17. Detail of washstand

11. 主卧
12. 客房
13. 主卧床头陈设
14. 主卧座椅
15. 浴室盥洗台
16. 浴室墙面采用石材
17. 盥洗盆特写

Neo-Classical Art in the New Century

The appearance of modernism has overturned the traditional design approaches in classicism. The concepts of simplicity and functionality have gained their popularity in interior design since the beginning of the 20th century. Following modernism we have the evolving trends like post-modernism, deconstructivism and surrealism. However, neo-classicism never loses its popularity. It has been, till today, a preference for interior design in luxurious hotels and grand villas. Nevertheless, neo-classical style in modern interiors is quite different from that in the 18th and 19th century. Old houses are adapted to new contexts and surroundings; new materials and emerging techniques change traditional approaches to neo-classicism; modern design concepts challenge the neo-

面向新世纪的新古典风格

现代主义的出现颠覆了传统的古典主义手法，以其简洁实用的理念占据了20世纪以来的室内设计风格。在其之后又出现了后现代主义、解构主义、超现实主义等各种流派，但新古典风格的地位从未因此动摇，时至今日，它依然是豪华酒店、大型别墅等类型空间的首选，只是在当今环境下的新古典风格已经与18、19世纪有了很大区别。古老的住宅在被新的环境改变；层出不穷的新材料、新技术改变着新古典风格的传统模式；现代设计理念对新古典的装饰效果提出了挑战；智能家居渐渐成为了新时代家居的主题；不同家庭的不同需求需要被满足……

越来越多的新形式让传统的新古典风格注定面临诸多挑战，也为其发展提供了更多的可能。对坚持这种风格的设计者来说，主动地融

CHAPTER 5
NEO-CLASSICAL ART IN MODERN HOMES
新古典家居的新形式

classical decoration; the idea of "smart home" becomes the theme of contemporary residences; particular requirements from different families should be considered in a personal way...

These new problems in modern homes have posed a big challenge for traditional neo-classical style, but at the same time, they have created opportunities for the development of neo-classicism in new circumstances. For those designers who keep to neo-classicism in their design, facing up to the new situation and finding solutions positively are the best choice. Only in this way, neo-classical style could be integrated into the life of modern man, instead of being a symbol for nostalgia.

合这些新形式，而不是被动的接受是唯一可走的路，也只有这样，新古典风格才不只是人们用来怀念旧时代的标志，而是真正融入到了现代人的生活中。

1. Renovation of Old Houses

In modern urban constructions, more and more attention is paid to the preservation of historic architecture. Meanwhile, to meet the need for comfortable living, renovation and restoration for the interior is necessary. Thus many spaces with combined or contrastive styles come into being, as well as various solutions to problems in renovation. Neo-classical style enjoys a long history and goes well with the façade of historic buildings, thus being a popular choice for such renovations. At the same time, with the advancement of social life and technology, new techniques, craftsmanship and materials are emerging and they have greatly changed the way interior design is approached.

In renovation projects, the most important part is the renovation

1. 旧环境的再创造

在如今的城市建设中，对古建筑的保护越来越受到重视。但为了适合人的居住，室内空间的翻新和修复却不能缺少，于是产生了各种风格交叉的复合空间，也派生出了对旧建筑改造的各种手法。新古典风格本身具有悠久的历史，能与旧建筑的外观相和谐，是旧建筑空间改造中常用的风格。但是随着社会生活和科技的发展，新技术、新工艺、新材料层出不穷，这些材料和技术的发展也改变着室内设计的面貌。

对旧建筑空间的改造最重要的部分是结构的改造。现代家居的功能越来越多，越来越复杂，一般情况下，原有建筑的结构都不能满足所需，用现代技术对原有空间进行结构性的调整就变得很重要。在改造和调整中，既要对原建筑结构进行考核，保留其中有价值的部分，又要考虑经济、安全等因素。旧建筑在采光上一般较差，在改

of building structure. In modern homes we have multiple spaces and complicated spatial relationships, which, more often than not, the existing structure cannot deal with. So we need to adapt the structure with modern techniques. In such renovations, you should examine the existing structure, retain the valuable parts, and take into account various factors such as budget and security. Generally speaking, lighting conditions in old buildings are poor, and extra openings can be added in renovation; or you can put more attention on a new reasonable layout. Vertically, you can adapt the ceiling height. In renovation, new equipments, devices and wires would be added, and the ceiling could be lowered to accommodate these facilities. New elements would be added in the structural renovation, and the reservation or demolition of old elements should be determined by their usefulness to your design.

A good arrangement in a space is one of the primary concerns in renovation projects. Furniture and furnishings are the main elements we use for such arrangements, and they act as a link between space and man. As an organic part of the interior, they connect the old space and new lifestyle. Therefore, in the selection of form, colour, texture and material we should seek a balance between traditional aesthetics and modern life requirements. The forms should go with the characteristics of the old building, while the texture and proportion should be comfortable and intimate. More importantly, historic architecture would have a peculiar sense of era perceived through the remaining patterns, columns, plasters… In a neo-classical home design, these symbolic elements could be used in the interior, but it should be noted that the arrangement of the interior space should match the original style and even enhance it.

The project presented here is a residence dating back to the 17th century, situated near the Big Ben and House of Parliament in London, a location that enjoys both a long history and rich culture. Therefore, in the renovation the aim is set to ensure comfort and usability while paying homage to the local history. Many existing details in the architecture are retained, and new spaces and facilities are added. After renovation, the residence is equipped with modern amenities such as lifts and air-conditioning system, combined with neo-classical decorations with remarkable English characteristics. Moreover, in order to match up with the feature of the existing architecture and interior, all the renovated rooms are overwhelmed with a cream (or similar) colour palette, integrating the past and present into one.

造过程中可以额外增加采光部分，或者在功能区的布置上做更多考虑。而在纵向上，也可以将高度做适当调整。旧空间改造后会增加许多新的线路和设备，可以在改造中将顶棚降低，节约空间的同时也可以把新的设备隐藏其中。结构性的改造需要有新的元素增加，但也要对旧建筑的构件有所取舍，充分利用其合理的部分。

其次，空间的组织对旧环境的改造也很重要。空间的组织主要是靠家具和陈设品，它们是环境与人之间沟通的媒介，也是室内环境中的有机组成，它们连接着旧建筑形态和新的生活方式，所以，在造型、色彩、材质等方面的选择应该兼顾到传统与现代两个方面，不仅形态能配合旧建筑空间的独特气氛，在形式和尺寸上也能让人感到舒适和亲切。更为重要的是，旧建筑中会带有自己独特的时代感，有时也会保留下原有的图案、柱式等组成部分，对新古典风格来说，这些具有象征意义的部分可以直接用到室内环境中，但在空间的组织中需要特别的衬托，以丰富空间的形象，并且体现出旧建筑空间原有的风格。

本案例是伦敦一处可以追溯到17世纪的住宅，并且在位置上靠近大本钟和国会大厦，既有悠久的历史，也有丰富的人文环境。所以对这处住宅的改造需要更多考虑如何在保证舒适和适用的情况下尊重历史。设计师在改造过程中保留了许多原建筑的细节，并且增加了许多新的功能区和设备，改造后的空间拥有电梯、空调等自动化系统，也有具有英国特色的新古典内饰，现代技术隐藏在其中。并且，为了配合原建筑和室内空间的特点，改造后的房间都用了奶油色或相近的颜色，使"过去"和"现在"浑然一体。

1. Fireplace detail in Ivory Requiem
1. 象牙安魂住宅壁炉装饰特写

Ivory Requiem
象牙安魂住宅

The house dates back to the 17th century and is an important house in the area of Westminster, in Central London, and within close proximity to Big Ben and the Houses of Parliament.

Much of the period detailing was rescued and magnificently refurbished including a new oval "lantern" skylight in the period style over the staircase.

Modern technologies including a lift, air conditioning, and full home automation systems were sympathetically and discreetly integrated whilst respecting the historic status of the house.

Whilst it was indeed a wonderful project and a designer's dream in terms of the possibilities, the client was not prepared to accept much in the way of exciting colour schemes and all the rooms basically had to be cream and/or cream variations. The designers managed to get a little colour in as long as the client had cream with it!

The client was particularly keen on a very traditional scheme throughout and indeed this suited the house well. The designers created a beautiful mahogany panelled study/library room which was masculine and classical. They had some fun on the lower floors where a swimming pool room and spa area was devised, and also designed a series of beautiful exterior terrace areas.

Location London, UK
Designer Stephen Ryan Design & Decoration
Photographer James Balston
Area 10,000m²

项目地址 英国，伦敦
设计师 史蒂芬莱恩设计装饰公司
摄影师 詹姆士·巴尔斯顿
项目面积 10,000 平方米

246 | NEO-CLASSICAL ART IN MODERN HOMES

这幢房屋可以追溯到17世纪，是威斯敏斯特地区很重要的房屋，位于伦敦中心，临近大本钟和英国国会大厦。

那一时期的设计细节大部分都经过了修复和重新翻新，显得富丽堂皇，并且在楼梯上方新增加了一个17世纪风格的椭圆形的灯笼式天窗。

一些现代技术，包括电梯、空调和全套的家庭自动化系统的综合应用，都充分考虑了该房屋的历史地位。

这确实是一个非常好的项目，从其发展可能性上来说，这也是设计师梦寐以求的项目。客户不接受令人兴奋的色彩方案，所有的房间基本上都采用了奶油色及其渐变色。设计师设法只加入一点点颜色，达到最后还是奶油色的整体效果。

客户自始至终都特别热衷于非常传统的设计方案，这确实也非常适合这幢房屋。设计师设计了一个非常漂亮的书房/图书室，其中运用了红木镶框的设计手法，具有浓郁的古典特色。他们还在低层中设计了一些娱乐活动场所，包括游泳池和水疗区，也设计了一系列非常漂亮的室外露台区。

2. Landing area
3. Entrance hall
4. Stairs in the hallway
5. Landing area on the ground floor
6. Kitchen decorated with wood
7. Dining table in the kitchen
8. Dining room
9. Classical living room
10. Cream living room
11. Bedroom for guests
12. Master bathroom
13. Bathroom vanity
14. Master bedroom

2. 楼梯平台
3. 入口门厅
4. 门厅里的楼梯
5. 一楼入口处
6. 厨房的材料采用木材
7. 厨房里的餐桌
8. 饭厅
9. 古典风格的客厅
10. 客厅采用奶油色色调
11. 客房
12. 主浴室
13. 浴室盥洗台
14. 主卧

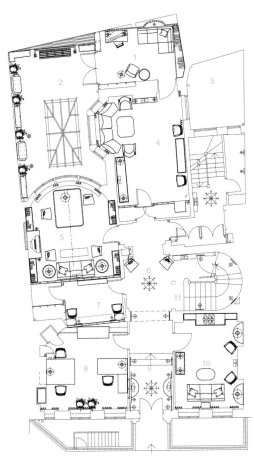

Ground Floor Plan
1. Library
2. Roof garden
3. Garden well
4. TV room
5. Study
6. Entrance lobby
7. Guest bathroom
8. Office
9. Entrance
10. Guest reception
11. Stairs

一楼平面图
1. 图书室
2. 屋顶花园
3. 花园采光井
4. 电视间
5. 书房
6. 入口门厅
7. 客用卫生间
8. 办公室
9. 入口
10. 待客室
11. 楼梯

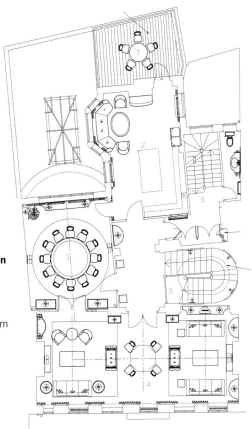

First Floor Plan
1. Terrace
2. Kitchen
3. Dining room
4. Drawing room
5. Stairs

二楼平面图
1. 露台
2. 厨房
3. 餐室
4. 客厅
5. 楼梯

2. New Design Languages for Old Styles

Nowadays with multi-culture societies and the advancement of information technology, trends for interior design evolve constantly, and various styles co-exist and influence each other. Although simplicity is the main stream in current interior designs, traditional neo-classical style still gains its popularity among architects, designers and their clients. In the contemporary context, neo-classical style tends to get simplified and abstracted, with many new elements added.

Firstly, lots of new materials are used in neo-classical design. Apart from consideration on aesthetics and durability, impacts on our health and the environment are the main factors for the selection of materials. In modern interior design, many new materials are

2. 用现代语言诠释传统模式

当今社会多元文化并行，信息技术日益发达，室内设计的风格流派也变化频繁，多种风格并存并相互影响，虽然简约是如今室内设计的主流，但传统的新古典风格还是受到很多人的喜爱。只是在新的社会背景和文化背景下，新古典风格有了明显的简化和抽象化，同时也加入了许多新的元素。

首先是越来越多的新材料的加入。除了考虑材料的美观性和耐用性之外，对人体健康和环境有无影响是目前选择材料的主要因素。在现代的室内装饰中，许多材料可以替代木材、石材等传统材料，并且可以产生很好的效果。塑料装饰板是一种新型的装饰材料，由高分子材料制成，可以替代传统的木材、钢材，并且它的质量轻、隔音、防火，安装方便，生产过程节能环保，应用范围很广，能满足不同场合的需要。微晶石也是目前应用广泛的一种装饰材料，

available to replace traditional ones such as wood and stone. Plastic boards are a good choice, for example. Made of polymer materials, they can be used to replace wood and steel; besides, they have advantages including light weight, sound and fire proofing, easy installation, environment-friendly manufacture process, and wide range of application for various requirements. Microlite is another material widely used in modern interior design. Natural marbles commonly used in neo-classical interiors would sometimes contain radioactive elements that have a bad effect on human health, and microlite provides an alternative. Made of natural inorganic materials, it has advantages such as natural, soft textures; multiple choices in colour; sustainability; and being economical (much cheaper than natural marbles). As for materials for walls, wallpapers and cloth are commonly used in neo-classical style designs. Decoration of walls is an important part in neo-classical style. Traditionally wooden boards would be used on walls, while nowadays, various kinds of wallpaper and cloth can be applied to present the appearance of traditional materials. In recent years, healthy and eco-friendly wallpapers were developed, evoking a revolution for non-toxic wallpapers. Now, convenient and inexpensive wallpapers and cloth have replaced the traditional material on wall – wooden boards – to provide multiple choices for your walls. As for fabrics, a wide range of application of new fabrics, final treatments, and the use of computer technologies provide opportunities for innovation in creating textile experiences. Some fabrics are given extra values beside their basic functions; for example, some curtains are good in sound-proofing and thermal insulation. The application of various new materials has changed the traditional approach to neo-classicism. The previously used materials with a high consumption on energy and non-renewable resources have been replaced while their advantages and characteristics are retained, showcasing a sustainable way of development.

In this project, the client envisioned both a traditional ambiance and cosy, contemporary living spaces for his apartment. Accordingly, some environment-friendly new materials are chosen, being durable and kept within the budget, among which the fabrics are work of a contemporary textile master. On the selection of furniture, pieces with both traditional and modern styles and a typical American touch are chosen; traditional profiles and cutting-edge styles are combined into one. The whole interior is overwhelmed by contrast between old and new, in which a new approach to neo-classical style finds its way.

新古典风格中常用的天然大理石有时会含有放射性元素，对人体产生影响，而微晶石采用天然无机材料，具有自然柔和的质感，色彩丰富，绿色环保，并且比天然大理石更经济。在墙面材料上，壁纸和墙布在新古典风格中经常会用到。墙面装饰是新古典风格的重要内容，传统的墙面装饰一般会用木制壁板，而如今，丰富多样的壁纸、墙布可以仿制出许多传统材料的外观，并且近年各种健康环保壁纸发展迅速，实现了壁纸无毒无害的革命性转变，更方便更经济的壁纸和墙布已经可以替代传统的木材实现墙面的多种装饰效果。在软装饰上，新织物纤维的开放使用、材料的后处理和计算机技术的应用也为室内装饰织物的创新提供了广阔的领域，一些织物除了传统的功能之外还具备了其他附加价值，例如，窗帘的隔音、隔热。诸多新材料的应用改变了新古典风格的传统形式，取代了原本能耗高、资源短缺的材料，又保留了它原有的特点，是其发展的新方向。

本案例中的业主想让自己的公寓有传统的氛围，而在繁忙的生活中又有舒适的现代环境。所以，设计师选择了一些环保型的新材料，考虑了预算的同时也经久耐用，其中的织物还出自现代纺织大师之手。在家具上又选择了结合传统和现代并且具有美国特色的样式，看上去既有传统的轮廓，又有前卫时尚的风格。整个环境充满了新与旧的碰撞和对比，在这种对比中也为新古典风格开辟了新的道路。

1. Dining room in Astor Apartment

1. 阿斯特公寓饭厅

Astor Apartment
阿斯特公寓

Working in one of Sydney's most iconic buildings, the 1920s Deco Regency, "Astor" on Macquarie St, Greg Natale Design wanted to create an updated apartment, fitting the grandness and prestige of the address. The client wanted to maintain a traditional flavour yet create a very functional modern apartment suitable for a busy city lifestyle. The client didn't want a nostalgic response but something that went with the vintage of the building and complemented the original interior, so Greg Natale chose to juxtapose the traditional with modern textiles by Kelly Wearstler, wallpapers by the legendary late English David Hicks and ceramics by Ettore Sottsass to give the apartment a modern edge.

The apartment was painstakingly restored with many original features retained. The Astor has a great policy for recycling architectural elements from their apartments so what was removed was used in other areas and what wasn't used in the apartment was given to the building management for other apartment refurbishments, including cornicing, skirtings, doors and floorboards. Because of this vast pool that the building has, few new materials were needed, which was positive in a sustainability context as well as budget. Low-VOC painting finishes, including the hand-painted joinery finish, were used throughout. The materials are highly luxurious yet highly durable to stand the test of time for another century.

Another important element used to create a contemporary interior was colour. The designers chose clean contemporary colours like dove grey, white, black, sage green and lilac. Using beiges and creams would have made the interior feel "twee" but using these shades helped give the apartment the harder edge desired.

The interior is the right mix of old and new, where old world glamour is teamed with hard-edged furniture. The furniture choice was another important element, where clean modern pieces by Rodolfo Dordoni for Minotti were mixed with 1940s "traditional" pieces by legendary American decorator Dorothy Draper. Working in a heritage listed A-grade building the designers successfully melded location appropriate styles for modern-day use. Rather than creating a modern box inside a traditional building, they chose to work with what they had and create an interior that is stylish but will age gracefully, like the envelope itself.

By using eclecticism to counteract what could have been a museum-like renovation, the interior decoration remains bold yet still complementary. The apartment features a strong sense of contrast both conceptually (old to new) and literally, seen in the strong pattern and tonal contrast. Both bedrooms include strongly patterned wallpapers to add individual interest, again, modern interpretations of classic patterns. The furnishings throughout are a mix of traditional silhouettes; adapted to be more fashion-forward and mid century and modern styles.

Location Sydney, Australia
Designer Greg Natale Design
Photographer Anson Smart
Area 180m²
Completion Date 2011

项目地址 澳大利亚，悉尼
设计师 格雷格·纳塔莱设计工作室
摄影师 安森·斯玛特
项目面积 180平方米
完成时间 2011年

阿斯特公寓位于麦格理大街上的一座古老建筑中，始建于20世纪20年代，悉尼最具代表性的建筑物之一。格雷格·纳塔莱设计工作室希望打造一间现代化的公寓，同时又要与其所在环境的恢弘大气相符。委托客户希望保留建筑的传统风情，同时要满足现代化公寓的使用功能，适合繁忙的都市生活节奏。客户不想要简单的怀旧风，而是要与这栋建筑物的复古风相融，在原来的室内装饰基础上有所改进，于是，设计师选择了传统与现代并置的手法，现代材料采用了凯利·惠尔斯特尔设计的织物、已故著名英国设计师大卫·希克斯设计的墙纸以及埃塔·索特萨斯设计的瓷砖，赋予这间公寓浓厚的现代气息。

Floor Plan
平面图

设计师煞费苦心地保留了公寓里原有的许多装饰元素。这栋大楼有一项规定，即公寓装修中拆卸的建筑材料要回收利用，从这里拆下的材料可以用在别处，没用上的材料交给大楼管理处，用于其他公寓的装修，包括檐口、壁脚板、门和地板等。由于有这项传统，所以装修中几乎不需要多少新材料，这不仅有利于环境可持续发展，而且能够节约开支。室内采用的装饰性涂料（如细木工制品的手工上漆）都是含低挥发性有机化合物的材料。装饰材料看上去极尽奢华，同时又能保证经久耐用，能够经受接下来一个世纪时间的考验。

为打造现代化的公寓室内空间而采用的另一个重要元素是色彩。设计师选择了干净简洁的现代色彩，如鸽子灰、白色、黑色、灰绿色和淡紫色等。米黄色和乳清色本来可能令室内看起来过于俗气，但是上述色调的使用却让公寓室内空间立刻现出棱角，令人眼前一亮。

室内空间是新与旧的混合，复古风情的空间搭配了棱角分明的现代家具。家具的选择是设计中的另一重点，选用了意大利设计大师鲁道夫·多多尼设计的米诺蒂系列家具，简洁、现代，搭配美国传奇装饰大师多萝西·德雷珀的传统风格作品，带来20世纪40年代特有的复古风情。在这栋A级历史保护建筑中，设计师成功融合了传统风格与现代用途。不是传统外壳与现代内核的简单叠加，而是在既定条件的基础上打造独特的室内空间，既前卫，又能随着时间的流逝愈显风情，就像这栋建筑一样。

2. Living room detail	6. The living room mixes old and new.	2. 客厅特写	6. 客厅空间是新与旧的混合
3. Entrance hall	7. Modern textiles by Kelly Wearstler used in the living room	3. 入口门厅	7. 客厅采用凯利·惠尔斯特尔设计的现代风格织物
4. Dining room		4. 饭厅	
5. Sage green kitchen	8. A corner of the living room	5. 厨房采用灰绿色色调	8. 客厅一角

设计师采用了兼收并蓄的设计手法，避免了出现像博物馆一样的室内装饰风格，室内装饰大胆而又现代。这间公寓体现出鲜明的反差，既有抽象的设计理念上的反差（新与旧的反差），又有视觉外观上实体的反差，体现在图案和色调的鲜明对比上。两间卧室都采用了图案鲜明的墙纸，凸显个性，同时这些图案也是对古典花纹的现代诠释。室内陈设采用了各种传统造型，经过改造，有的更具时尚感，有的呈现出中世纪风格，有的则变成现代风格。

9. The bedroom includes strongly patterned wallpapers.
10. The bedroom is filled with hard-edged furniture.
11. Detail of bedroom
12. The bathroom decorated with ceramics by Ettore Sottsass

9. 卧室墙纸的花纹非常抢眼
10. 卧室里的家具棱角分明
11. 卧室特写
12. 浴室采用埃塔·索特萨斯设计的瓷砖

3. Modern Design Concepts in Neo-Classical Homes

Neo-classicism is an open and inclusive style. Now, modern concepts such as individualism and minimalism also contribute to the development of neo-classicism. Individuality and simplicity are reflected in neo-classicism as the abstraction, simplification and integration of traditional styles. The garish and pompous decorations are stripped off from neo-classicism, which is thus simplified into artistic symbols with pure colours. Sometimes elements with other styles would be used to add a peculiar character. However, simplicity doesn't mean lack of design while individuality is not deprived of the very spirit of neo-classicism. The cultural background and pursuing of spirit in neo-classicism should be the values that we always stick to.

3. 新古典家居中的现代设计理念

新古典风格本身具有开放包容的精神，如今，个性化和简约化的现代设计理念也影响着新古典风格的发展。现代设计对简约、个性的强调在新古典风格中表现为对传统的抽象、简化和融合。传统装饰去掉了那些华丽、浮夸的内容，被简化为一些符号，色彩上也更纯粹，有时也会直接引用其他风格的元素以示独特。但简约并不意味着什么都没有，个性化也不意味着脱离新古典的内涵，新古典风格的文化背景、精神诉求依然是设计中应有的坚持。

对传统的简化和抽象通常集中在新古典风格常用的几个元素上：柱式、壁炉、墙面装饰、拱券形式。柱式在室内的应用一般只是装饰，古典柱式本身有严格的比例要求和形式要求，但在室内装饰中，有时也会根据需要改变它的造型，或者简化为壁柱，装饰在壁炉两侧或墙面上。壁炉在室内处于中心位置，所以比较注重装饰

Simplification and abstraction of tradition are realised commonly with several elements in neo-classical style, including columns, fireplaces, wall decorations and arches. Interior columns serve as a kind of decoration apart from their architectural functions. For traditional columns there are strict requirements on forms and proportions, but in interior design, we can make a bit of adaptation. For example, sometimes interior columns can be simplified into pilasters as decorations on walls or on both sides of the fireplace. The fireplace plays a central role in an interior space and is often the focus of interior decoration. However, sometimes, particularly when the fireplace exists in the bedroom, trivial decorative elements would be taken out to present a simple, contemporary image. As for decoration for walls or doors and windows, simple and clear moldings are often used in modern interior design; sometimes wallpapers would be adopted to save troubles.

Concerning the colour palette, red, gold and yellow are most frequently used, which are rarely employed in an extensive way in contemporary designs. Usually light colours can be used in combination with grey to reveal a contemporary sense in neo-classical style. Particularly for decoration of walls, which occupy a large proportion in the interior surface area and play the role of backdrop for furniture and furnishings, the colours selected would set the basic tone of a room. Therefore, wall colours are one of the primary concerns in creating interior atmospheres. Influenced by modern design, now the neo-classical style interior is often deprived of the golden inlay parts, replaced by simple and pure colours. Reduction in the use of colours would make a space simple and clean. Meanwhile, we should pay more attention on textures of materials, craftsmanship and the selection of furnishings to re-establish the dignity and gracefulness that belong to neo-classicism in nature.

The project selected is a modern house with state-of-the-art facilities and the best materials. Here modernity is combined with a bit of neo-classicism. Many existing neo-classical elements are retained, such as the marble fireplace, detailed moldings, and arch windows. On the whole there is neither overstated decoration nor radiant colours, but the glory of neo-classicism is subtly revealed with the rhythmic patterns intricately carved in the white ceiling plaster and the wall panels in the same colour palette. There is no fancy furniture or showy fabrics, but a luxurious atmosphere is created. The traditional elements are used just in moderation, while a strong impression is established. You can hardly say it belongs to certain styles; or, we'd rather say that modern concepts have been well integrated into neo-classical spaces.

性，但有时也会根据情况去掉繁琐的装饰，尤其是在卧室中，常常会采用更现代更简洁的壁炉。而对墙面或门窗等位置的装饰来说，现代的室内设计中常常会用更简洁的线条来勾勒造型，或者直接用墙纸替代。

在色彩上，新古典风格常用红色、金色、黄色等，而现代设计中这些颜色较少大面积的使用。通常，用高明度的颜色掺杂灰色可以在新古典风格中体现出时代感。尤其是在墙面的装饰上，墙面色彩在家居装饰中所占面积最大，并且充当家具、陈设的背景色和室内色彩的基调，所以，墙面色彩的选择对家居情调的营造非常重要。受现代设计的影响，过去新古典风格中的金色镶嵌常常会省略，取而代之的是更简单纯粹的色彩，色彩的减少能使空间显得更简洁，但在室内材料的材质、工艺以及其他陈设上需要更注重肌理效果，以显出新古典风格固有的庄重和华丽。

本案例是一所现代化的住宅，室内拥有最先进的设施和最好的材料。设计风格在现代的基础上也保留了新古典风格中常见的元素，大理石壁炉、精心雕刻的线脚、拱形窗户。但在整体上没有过多的装饰和绚丽的色彩，白色的石膏天花板雕刻着有规律的线条，墙面也是同色的护墙板。家具和织物并不花哨，但都能体现出华丽的质感，在低调中能感受到奢华的气氛。古典元素在这里点到即止，但依然让人印象深刻，让人很难判断它到底属于何种风格，或者可以说，这是现代设计的理念深深印在了有新古典风格的空间中。

1. Guest suite in Down Street Apartment

1. 唐郡街公寓客房

Down Street Apartment
唐郡街公寓

A rare opportunity to acquire a newly modernised five-bedroom lateral apartment extending to approximately 355sqm. With two intercommunicating reception rooms opening onto a balcony and accessed off the generous entrance hall, it is ideally suited for entertaining on any scale. The bedrooms are discreetly separated from the reception areas, in particular the master suite which has its own reception room as well as a generous dressing room and en-suite bathroom.

The apartment has been comprehensively redeveloped by leading designers Casa Forma incorporating the latest technology and the finest materials. It has been interior designed and furnished, showcasing unique furniture and innovative finishes.

A spacious and impressive separate entrance hallway retaining many traditional features, updated with contemporary details and fittings. It has original Edwardian strap-work plaster ceiling and cornice, original wainscot panelling which is restored and has a contemporary painted finish, and existing leaded window embellished with bespoke stained glass. Floor in Statuario Lucido white marble set within a timber border, has under-floor heating.

There are also Venetian glass chandelier and mirror, and bespoke joinery storage cupboard.

Leading directly from the entrance hall, the reception room is a grand space flooded with light from the full-length French Doors set into a bay. Fully furnished, it features separate drinks or bar area leading to a dedicated audio visual storage area and new inlaid polished timber flooring with a luxury silk area rug. The original French Doors to bay are restored, and dressed with motorised blinds; sheer and feature drapes are also included.

The dining room provides a second spacious entertaining space. Original fireplace restored to working order with a gas-fired system. Original cornice, picture rails and feature mouldings restored or reinstated.

Discreetly located behind the main entrance hall, the guest cloakroom provides luxurious guest WC facilities with Belgium black marble vanity, gilded glass basin and bespoke joinery, new timber flooring and bespoke leaded stain glass window.

Location London, UK
Designer Casa Forma Limited
Photographer Casa Forma Limited
Area 355m²

项目地址 英国，伦敦
设计师 家居形式设计公司
摄影师 家居形式设计公司
项目面积 355平方米

Lobby Level Plan

1. Entrance hall
2. Reception room
3. Bar area
4. AV area
5. Dining room
6. Guest W.C.
7. Kitchen
8. En-suite bathroom
9. Bedroom
10. Master bedroom study
11. Walk-in wardrobe
12. Master bedroom
13. Master bathroom
14. Utility room
15. Maid's bedroom
16. Guest en-suite

大堂平面图

1. 入口大厅
2. 待客室
3. 吧台
4. 多媒体设备
5. 餐室
6. 客用卫生间
7. 厨房
8. 套房卫生间
9. 卧室
10. 主卧室里的书房
11. 步入式衣橱
12. 主卧室
13. 主卫生间
14. 设备间
15. 女仆卧室
16. 客用套房

这套公寓有5间卧室，经过扩建，面积达到了355平方米，配备了现代化的设施。两间会客室从内部相连，毗邻阳台，从宽敞的入口门厅可以直接进入会客室，非常适合各种规模的聚会宴请活动。卧室特地设置在远离会客区的位置，尤其是主卧，主卧除了有自己单独的会客室，还有宽敞的更衣室和浴室。

伦敦知名的家居形式设计公司对这套公寓进行了彻底的翻修改造，采用了最新的技术、最好的材料。除了室内的设计和装修之外，还包括独特家具的选用和创意的装饰。

独立的入口门厅非常宽敞，令人过目难忘，里面保留了许多传统装饰元素，利用现代化的配件和细部装饰进行了修饰。这里有爱德华七世时代原始的灰泥吊顶和檐口，有最原始的护壁板（经过修复，表面采用了现代化的装饰涂料），还有古老的开窗，采用定制的彩色玻璃进行装饰。地面采用白色大理石，四周有木质镶框，采用地热。吊灯和镜面采用威尼斯彩色装饰玻璃，储藏柜是定制的细木工制品。

从入口门厅进入会客室，你会发现后者也一样宽敞大气，法式落地双扇玻璃门为室内带来充足的光线。会客室里有着丰富的装饰陈设，有独立的饮酒区，旁边是专门的视听设备存放区，地面采用嵌入式抛光木质地板，铺了一张奢华的丝质小地毯。古老的法式双扇门经过修复，增加了机械化的百叶窗帘，此外还有透明的特色窗帘。

2. Generous entrance hall
3. Bar cabinet
4. Long corridor
5. Spacious kitchen
6. The dining room provides a second spacious entertaining space.
7. Dining room detail
8. Dining room fireplace
9. The reception room is a grand space.
10. Study

2. 宽敞大气的入口门厅
3. 储藏柜
4. 长长的走廊
5. 宽敞的厨房
6. 饭厅提供了另外一处宽敞的宴请场地
7. 饭厅一瞥
8. 饭厅壁炉特写
9. 会客室宽敞明亮
10. 书房

饭厅提供了另外一处宽敞的宴请场地。古老的壁炉经过修复，采用了现代的燃气系统。原来的檐口、挂镜线和独具特色的装饰性线条都进行了复原。

主入口门厅后面就是宾客衣帽间，这里有奢华的客用卫生间，采用比利时黑色大理石、镀金玻璃洗手盆以及各种精细木工制品，新铺了木质地板，安装了定制的彩色玻璃窗。

11. Guest suite
12. Bedroom of guest suite
13. Study area in master suite
14. Master bathroom
15. Detail of master bedroom
16. Chair in master bedroom
17. Guest cloakroom

11. 客用套房
12. 客用套房的卧室
13. 主套房里的学习空间
14. 主浴室
15. 主卧陈设特写
16. 主卧桌椅
17. 宾客衣帽间

4. Smart Homes in Neo-Classical Style

Modern information technology, with the advancement of computers and networks, has greatly improved social developments, and also made its contribution to smart homes. Nowadays, any house, no matter designed in what styles, cannot go without various electric appliances. A smart home that celebrates the integration of network communication, information appliances, and facility automation indicates the future trend of home design, in which efficiency, safety, comfort, convenience and sustainability are all involved.

A smart home is one that uses computer technology, network communication technology and structured cabling system to connect sub-systems related to home life in an organic way. Hence, compared with traditional home spaces, a smart home is not a passive, static

4. 智能新古典家居

以计算机和现代网络技术为特征的现代信息技术促进了社会的发展，在家居环境中也发挥了巨大的作用。如今的家居环境中，无论何种风格，都离不开各种电器以及其他设备的控制。而以住宅为平台，兼备网络通信、信息家电、设备自动化为一体的高效、舒适、安全、便利、环保的智能化家居更是未来家居发展的方向。

智能家居是利用计算机技术、网络通信技术、综合布线技术将与家居生活有关的各子系统有机地结合起来，因此，与传统的家居环境相比，智能家居并不是被动的静止结构，而是有能动性的工具，可以有效地优化人们的生活方式。目前，智能家居由于造价等原因并没有普及，一般只适合于中高档的住宅。

目前的智能家居系统常见的配置有中央控制系统、智能窗帘系统、

structure, but a dynamic tool to efficiently optimise our lifestyle. Currently due to the reason of a high cost, smart home systems are not popular among the general public households, and are mainly adopted in middle and top grade residences.

The present smart home systems are usually equipped with central control system, intelligent curtain system, intelligent light system, intelligent household appliances system, background music system, security and protection system, access control system, remote control system, etc. Besides these, some auxiliary systems can be applied, for example equipments for pets, intelligent bath system, and automatic irrigation system. You can choose several of them according to your needs, and connect them to a central control system. There are options for the size and shape of the hardware equipments to match different interior design styles, but sometimes certain equipments just cannot be adapted, and flexible solutions would be needed. For neo-classical style, the design of walls and ceilings is particularly important. Lighting devices, loudspeakers of the music system, air outlets of the air-conditioner, household appliances that can be hanged or mounted on the wall… All these can be visual annoyances for a neat space and should be hidden or integrated with proper treatments of the surface. Meanwhile, these high-speed, high-efficiency equipments should match up with the neo-classical style in terms of sound, light and colour to avoid any inconsistency in the whole interior. The aim of applying smart home systems is to make life easier and more comfortable; home spaces shouldn't be crowded or made noisy with equipments and machines. It is necessary to think carefully about the effects of the systems and take the design concept into account. All in all, systems are used to serve men.

The selected project is a luxurious house with approximately 1,000sqm on three floors. Apart from living spaces, there are recreational rooms such as a billiard room, a game room, and a gym. The house enjoys a good location with beautiful natural sceneries with mountains and water outside, and a unique interior style is thus created to match the outdoor landscape. Decorated in a complicated yet orderly way, each room follows the principles of neo-classical style. Even the exterior landscape is designed in a symmetrical manner. It is a traditional looking house, but the interior is equipped with all high-end systems including lighting, music, curtain and household appliances, all connected to a central control system, completing an integrated smart home. Most importantly, the overall style is not disturbed by the intelligent systems; neo-classical aesthetics and the comforts of contemporary life are perfectly combined.

智能照明系统、智能家电系统、背景音乐系统、安防系统、门禁系统、远程控制系统。另外还有一些辅助的系统，如宠物设备、智能卫浴系统、自动浇花系统等。这些系统可以按需求选择几种，由中央控制系统集中控制。智能系统的一些硬件设备有不同的外形，可以适应室内的设计风格，而对一些无法改变外观的设备来说，就需要在设计中综合考虑。新古典风格本身就很注重墙面和天棚的设计，所以，照明系统的灯具、音乐系统的扬声器、空调的出风口、一些可以悬吊或靠墙的家电都可以通过界面本身的装饰隐藏或弱化。同时，这些高速度高效率的设备在声音、光影、色彩上也要与新古典风格相匹配，否则会在家居中产生不协调感。智能家居系统的目的是为了创造更舒适和人性化的环境，不能使家居空间变成充满机器的嘈杂环境，所以，在使用之前就需要对智能系统做出选择，并且充分考虑家居空间的设计风格，使系统能真正地为人服务。

本案例是一所近1000平方米的豪华住宅，三个楼层除了居住空间之外还包括了台球室、游戏室、健身房等休闲空间。住宅的位置依山傍水、风景秀丽，所以设计师也特意在室内创造出了独特的风格。虽然室内的装饰复杂，但每一个空间都有条不紊，遵循着新古典风格的原则。即使是室外的景观也设计成了对称的形式。尽管造型传统，但住宅内却配有所有高端的设备，灯光、音响、窗帘、电器设备等都被中央系统控制，可以称得上是一个完整的智能家居。并且，智能化的设备并没有影响环境的设计风格，传统的美感和现代的舒适完美地结合在了一起。

1. Breakfast patio in Mike&Sheila Mokhtare Family Residence

1. 莫克特别墅早餐庭院

Mike & Sheila Mokhtare Family Residence
莫克特别墅

This opulent custom estate is located in the heart of Newport Coast, in a prestigious gated community in Orange County, California. This house is comfortably nestled on the hill side with magnificent sweeping views of the Pacific Ocean during the day and breathtaking glimmer of city lights from the nearby Lido and Balboa Island on the evenings. One could feel lost in the immaculate views and feel as if they are swept away in a European city off the Mediterranean Sea.

On this project, Sean Hashemian, principal designer at Casa Siena, collaborated with the Architect, David Pierce Hohmann, from the beginning stages of conceptual design. Casa Siena was successful in creating a one-of-a-kind custom home that the discerning client, Mike and Sheila Mokhtare, had envisioned and dreamt of for their family residence. The house is composed of a sophisticated traditional architectural vocabulary. The designers have embedded their own unique touches within this realm of style to set the house apart from its neighbouring properties.

The house is roughly 929 square metres of livable areas, and encompasses three floors. No expenses were spared in the design and construction of this home. The house has a dramatic entrance with a double staircase, formal living and dining room, family room, breakfast dining room, kitchen and a secondary kitchen for staging and catering, six bedroom suites, six full bathrooms and two half bathrooms, library, media room, billiards room, wine cellar, game room, spa, gym, gardener's room, and outdoor loggia. The house is fully equipped with all the high-end amenities and technological gadgets of the 21st century making it a complete smart home. The house has full integration of lighting, audio, drapery, and mechanical control with the use of the Vantage system combined with the Crestron system.

All the tedious interior detailing has been completed by Casa Siena, which grouped some of the most talented people in the industry with their unique one-of-a-kind design to create a masterpiece for the client. Many of the materials and furniture items used in the house are unique to the house and were custom designed for this project. In this day and age where everyone's motto has become less is more, Casa Siena have stayed true to their traditional roots. They believe that more is more, but they aim to take the traditional to a different level by implementing and adding a more contemporary edge to it. They strive to come up with designs that embody the lifestyle of their clients and still have a new fresh outlook. Through the years they have truly mastered the look of the new traditional!

Location Newport Coast, California, USA
Designer Casa Siena / Sean Hashemian
Photographer Michael Garland Photography
Area 929m²

项目地址 美国，加利福尼亚州，新港海岸
设计师 锡耶纳家居设计公司（肖恩·哈什米安）
摄影师 迈克尔·加兰摄影工作室
项目面积 929平方米

这栋奢华的定制别墅位于加州新港海岸中心，在奥兰治县一个知名的封闭式社区内。这栋别墅坐落在山坡上，周围有极好的视野，白天能够遥望太平洋，傍晚能欣赏附近的丽都&巴波亚岛上灯光闪烁的美景。在如此完美景色的怀抱中，你会忘了身在何处，以为自己是在远离地中海的欧洲某城市中。

在本案的设计中，主持设计师肖恩·哈什米安从最初的设计理念开发阶段就与建筑师大卫·皮尔斯·霍曼合作。锡耶纳家居设计公司成功打造了一栋独一无二的别墅，使得品位出众的委托客户夫妇——麦克·莫克特和希拉·莫克特——实现了对他们的家庭住宅的幻想。这栋别墅采用精致的传统建筑语汇。设计师在其中融入了独特的设计手法，使其在周围房屋中脱颖而出。

这栋别墅的居住面积约为929平方米，共有三层，设计和施工的预算没有一点浪费。别墅入口对称布局的台阶恢弘大气，室内有起居室、家庭活动室、早餐室、厨房（包括主要和次要厨房）、6间套房、6个浴室、两个卫生间、图书室、多媒体室、台球室、酒窖、游戏室、SPA水疗室、健身室、园丁室以及户外凉廊。这栋别墅配备了全套的高档设备，采用了21世纪最先进的技术装置，真正实现了"智能家居"的生活方式。别墅采用"优势"智能家居定制系统（Vantage）和"快思聪"智能家居控制系统（Crestron），实现了照明、声音、窗帘以及各种机械设备的一体式操控。

室内繁重的细部工作也都是由锡耶纳家居设计公司完成的。该公司汇集了室内设计领域内最杰出的一批专业人士，为客户打造了独一无二的设计。室内所用的许多材料和家具都是为本案专门设计的。当今时代，似乎每个人都相信"少即是多"的理念，而锡耶纳家居设计公司却始终坚信"多才是多"。他们希望能够通过增加现代元素来赋予古典风格一个新的高度。锡耶纳的设计旨在针对委托客户的生活方式，满足他们的日常需求，同时又要让空间呈现出全新的面貌。通过多年设计经验的积累，锡耶纳已经掌握了"新古典"的本质。

2. Entry foyer
3. Family room on the ground floor
4. Warm family room
5. Formal living room
6. Main kitchen
7. Basement gardening room
8. Breakfast nook
9. Formal dining room
10. Ceiling detail in the dining room
11. Billiard in the basement
12. Traditional library
13. Wine cellar

2. 入口大厅
3. 一楼家庭活动室
4. 家庭活动室温馨明媚
5. 起居室
6. 主厨房
7. 地下室园艺室
8. 早餐室
9. 饭厅
10. 饭厅天花特写
11. 地下室台球室
12. 传统风格的图书室
13. 酒窖

Ground Floor Plan
一楼平面图

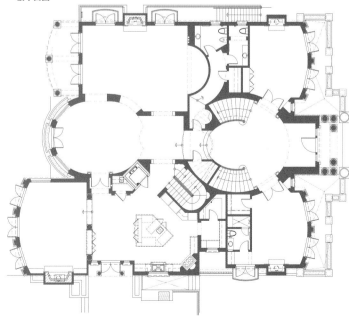

First Floor Plan
二楼平面图

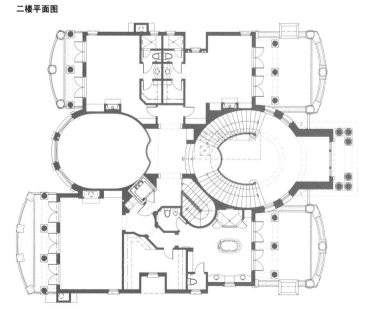

Basement Floor Plan
地下室平面图

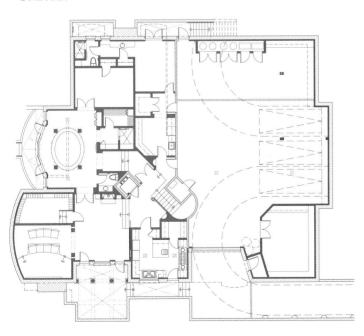

新古典家居的新形式 | 279

14. Daughter's bedroom
15. Guest room
16. Master bedroom
17. Men's powder
18. Women's powder

14. 女儿卧室
15. 客房
16. 主卧
17. 男士盥洗台
18. 女士盥洗台

5. Humanism and Sentiments in Home Design

When modern households have been transformed from traditional residences into smart homes, the power of technology has so greatly affected our life. Traditional concepts on a home life are overturned, and a new way of home living comes forth which highlights comfort, convenience and privacy. Yet the essence of this new home living lies not in getting rid of trivial housework or controlling various equipments; rather, it lies in meeting the most subtle requirements and desires with natural contexts and artificial means, and in this process certain sentiments would be established contributing to the creation of private and personal home settings.

Humanism in home design means the integration of nature, environment and man. In the process of design, we should

5. 家居设计的人性化和情感化

当现代家居环境开始由传统住宅转变为智能住宅时，科技的力量开始渗透进我们的生活，并颠覆了过去的家居生活理念，开始了一种全新的家居生活方式——安全、舒适、便捷。但新的生活方式并不仅仅是让人摆脱繁琐的家务或是对各种设备的控制，它的意义更在于设计中，用自然条件和人工手段把人性的最细微愿望和需求表现出来，并融入一定的情感因素，传达给他人，也就是设计中的人性化和情感化。

人性化的设计是自然、环境、人的和谐统一，在设计过程中，根据人的行为习惯、生理结构、心理情况、思维方式等，对原有的设计进行优化，在设计中体现人文关怀和对人的尊重。人性化的设计强调服务于人，但这并不意味着家居环境一定要使用大量的科技产品来取代人的活动，服务于人的前提是充分了解所服务对象的需求。

optimise the space according to the case of the inhabitant: the behavioural habit, physiological and psychological conditions, and the way of thinking... showing the humanistic care. Humanistic design highlights human orientation, but it doesn't mean lots of technological products are to be used to replace human activities. A prerequisite for human orientation is a full understanding of the requirement of the user. Each family has its individuality: members with different races, cultural backgrounds, aesthetic habits, and requirements for comfort... These are factors that determine the details in your design. Diversity of clients and their requirements makes research about each family member necessary for a designer. Only when their particular needs are identified could a designer set specific goals in a project.

Today with the fast development of science and technology, sentiments ironically become one of the main concerns in home design. We need to be served with hi-tech systems, but more importantly, cared with interests and sentiments. Sentiments play an important part in creating an interior space, home spaces in particular, where we expect more intimacy and personality. A home should be a place where we can feel pleasure and enjoyment when staying. For designers, the establishment of sentimental spaces involves the form, colour and texture of all the elements, plus sound, light and shadow, and even smell, to create a setting or an image that recalls the beautiful memories of the inhabitant. Neo-classicism is rooted in the rich European culture, which highlights the renaissance of classical art. The Christian history and culture as well as literature are usually the inspirations for the design of neo-classical sentimental interiors. Generally such elements would not be extensively used but only appear in decorative paintings and furnishings.

The project selected here is a special one – mansion of the Russian President. The castle was once used as an orphanage during the Soviet Union period. Before restoration, decorations from that period could still be seen in the interior. To meet the special needs of a mansion for the President, certain parts were added or reconstructed, and pieces of feature furniture and dining facilities were added. A special feature different from other housing is the inclusion of some hotel spaces for reception. After renovation, the mansion is characterised with wooden elements with a medieval style. From the furniture to decorative materials and fabrics, everything recalls the mysterious medieval castle. The functions are programmed for the needs of the special inhabitant. The humanistic and sentimental interiors made it one of the most respected government residences in Russia.

每个家庭都有自己的特殊性，家庭成员的种族、文化背景、审美习惯、不同人对舒适度的要求，这些都影响着设计中的细节，服务对象的多样决定了设计师需要对每个家庭成员进行研究，理解了他们的特殊需求，这样才能在设计中有针对性。

在技术不断发展的今天，情感化的设计是人们希望能在室内空间得到的更多情趣和关怀，满足更多的情感需要。情感是营造室内环境意境的重要因素，尤其是在家居环境中，人们都希望环境能更有人情味、更有情趣，置身其中，可以得到精神上的享受。对设计师来说，情感化空间的营造就是借助于室内各种元素的形状、线条、色彩，加上声音、光影、甚至味道去塑造一个个场景、画面，引起人们的联想和记忆。新古典风格本身建立在欧洲丰富的文化背景下，强调复兴古典艺术，对情感的把握也很有节制，常用基督教历史文化和文学性的主题形象作为情感的象征。对一般的家居空间来说，这些情感象征并不需要大量存在，通常只会出现在装饰性的油画、陈设等元素上。

本案例是一个极特殊的住宅——俄罗斯总统的官邸。这座城堡在苏联时期曾经是一座孤儿院，在设计师修复之前，室内还有一些当时留下的内饰。为了使修复后的住宅能满足总统住宅特殊的功能，设计师针对一些特殊功能做了增建和改造，增添了有特色的家具、餐饮设施，与其他住宅不同的是，还特别增加了一部分空间作为接待用的酒店。完成后的总统住宅以木制装饰元素为主，融合了中世纪的风格，从家具到装饰材料到织物的运用都让人联想到中世纪神秘的城堡，而在功能上，也符合居住者的身份及需求，人性化和情感化的设计让这个住宅独一无二，并成为了最受尊敬的政府住宅之一。

1. Hall in Meindorf Castle

1. 梅恩道夫城堡门厅

Meindorf
梅恩道夫城堡

DOM-A studio took part in the reconstruction and restoration project of Meiendorf Castle together with OAO (Open Joint-Stock Company) "Mosoblstroyrestavratsiya". Meiendorf Castle is the official residence of the president of Russia. Putin and Medvedev accepted there the President of France Sarkozy and the president of Azerbaijan Ildar Aliyev.

This project is unique, just by itself, as it is a unique location in Moscow, built in Gothic style, and the walls were saved by miracle: in Soviet period the building was used as an orphanage. The task the designers solved isn't less unique: even if in life of the architect such project will be met only once; it is already great luck. No more than 10 % of the original decor was remained: here a parquet scrap, there a fireplace, here a fabric fragment… and on these remains and on few remained archival materials it was necessary to restore the authentic shape of the building, to revive the interiors which were not saved till designer's time and ancient extensions. From the functionality point of view it was supposed to adapt an ancient mansion for accommodation under the house of receptions with a small hotel zone of the highest level.

For the solution of these tasks the reconstruction was combined with restoration works. With unique elements of decor and architecture the designers worked together with highly professional colleagues from Mosoblrestavratsiya. They added a number of new functions to the castle, redistributed the former (including a catering part, a hardware part, and a hotel part), and created difficult designs of the built-in furniture.

There was a difficult task for architects and decorators of the studio – to recreate internal space of the Castle, its spirit and identity in accurate compliance with historical drawings and photos. Three main halls were restored. In these halls maximum quantity of information and even furnish details till our time were saved, for example, a gobelin in a hall of receptions, fireplaces, and wooden decor elements. The Hunting fireplace in a living room enters the top ten most known fireplaces in the world. The main staircase of the Castle was repeatedly altered by owners and its image was created as a combination of elements from different decades, but all of them are harmoniously connected among themselves. Uniform design of interior, was carefully recreated by pieces of furniture, finishing materials, curtains and elements of decor in medieval castle style and adapted to new purpose of the Castle as the residence.

Location Moscow, Russia
Designer Dom-A/Maria Serebryanaya, Sergey Makushev
Photographer Aleksey Knyazev
Area 7,500m²

项目地址 俄罗斯，莫斯科
设计师 Dom-A工作室（玛利亚·赛瑞布莱恩娜雅，谢尔盖·马库谢夫）
摄影师 阿列克谢·肯亚泽夫
项目面积 7,500平方米

DOM-A工作室与来自《Mosoblstroyrestavratsiya》的OAO(开放合资公司)共同参与了梅恩道夫城堡的科学重建和修复工程。梅恩道夫城堡是俄罗斯总统的官邸，普京和梅德韦杰夫曾在那接待过法国总统萨科奇和阿塞拜疆总统埃达·阿利耶夫。

该项目本身就很独特，因为它位于莫斯科重要的区域，是一幢哥特式的建筑。建筑的墙壁都奇迹般地保存完好，苏维埃时期，这里曾被用作孤儿院。设计师所承接的任务也不乏独特性，设计师一生能遇到过一次就已经是无比幸运了。原先保留下来的装饰不足10%，有一些镶木地板废料、一个壁炉，还有一些纺织物碎片。设计师必须要在这些遗迹中，利用这些少有余留的材料，通过向历史时期的延伸，来恢复建筑的真正形状，还原室内空间的本色。从功能性的角度来看，应该在还原这座古老的宅邸的同时，增加其接待的功能，在顶层增加一个小型酒店区。

要将重建工作和修复工作相结合，共同完成这些任务。设计师与来自Mosoblrestavratsiya的非常专业的同行们共同努力完成这些独特的装饰和建筑元素。他们还给这座城堡重新分区，增加了一些新的功能区（包括一个餐饮区、一个五金器具区和一个酒店区），打造了一些困难的设计，如嵌壁式家具。

工作室的建筑师和装饰家们曾经遇到了一个非常困难的工作——重新打造的城堡的室内空间，在精神面貌和身份特征上要与历史遗存的图纸和照片完全相符。三个主厅经过了重新修复，这些大厅都保留了最多的原始信息和装饰细节，例如，大厅接待处中的仿哥白林挂毯、壁炉，还有一些木制装饰元素等。其中一个客厅中绘有狩猎图的壁炉已入选世界10大著名壁炉。城堡中的主楼梯经过主人们的反复改变，已经成为了代表不同时期元素的结合体，但是所有元素又可以相互融合。为了适应城堡作为住宅这个新的用途，设计师通过各种家具、饰面材料、窗帘和其他一些中世纪城堡风格的装饰元素的搭配使用，精心创造了室内的匀称设计。

288 | NEO-CLASSICAL ART IN MODERN HOMES

2. Chimney hall
3. Fireplace in the centre of chimney hall
4. Cabinet of the baron
5. The cabinet full of wooden decor
6. Alcove in the chimney hall
7. Stair on the first floor
8-9. Main staircase of the Castle
10. The staircase was repeatedly altered by owners.

2. "烟囱大厅"
3. "烟囱大厅"中央是个壁炉
4. "男爵大厅"
5. "男爵大厅"采用大量木质装饰元素
6. "烟囱大厅"凹室
7. 二楼楼梯
8、9. 城堡主楼梯
10. 城堡中的主楼梯经过主人们反复改变

11. Drawing room
12. Small drawing room on the first floor
13. Hunting hall
14. Banquet room

11. 会客厅
12. 二楼小会客室
13. "狩猎大厅"
14. 宴会室

15. Bedroom on the first floor
16. Medieval castle style bedroom

15. 二楼卧室
16. 卧室呈现出中世纪的城堡风格

Sketch of Bedroom
卧室手绘图

INDEX 索引

Baharev&Partners
Web: www.baharev.ru
Email: info@baharev.ru

Anna Kulikova & Pavel Mironov
Web:http://mironov-studio.ru
Email: mironov-studio@mail.ru

Dmitry Velikovsky
Web: http://dvelikovsky.com
Email: zg@artistic-design.ru

Dom-A
Web: http://a-dom-a.ru
Email: office@dom-a.ru

AS 20/10
Web: http://arst2010.ru
Email: info@arst2010.ru

Design-Bureau of Marina Putilovskaya
Web: http://putilovskaya.ru
Email: mpdesignbureau@mail.ru

Lauren Ostrow Interior Design, Inc.
Web: www.laurenostrow.com
Email: roomatologist@aol.com

Nico van der Meulen Architects
Web: www.nicovdmeulen.com
Email: nico@nicovdmeulen.com

Providence Ltd. Design
Web: http://providenceltddesign.com
Email: mona-thompson@att.net

Elia Felices interiorismo
Web: http://eliafelices.com
Email: press@eliafelices.com

Greg Natale Design
Web: www.gregnatale.com
Email: info@gregnatale.com

Landry Design Group
Web: www.landrydesigngroup.com
Email: contactus@landrydesign.net

S.B. Long Interiors
Web: http://sblonginteriors.com
Email: info@sblonginteriors.com

Phil Kean Designs
Web: http://philkeandesigns.com
Email: amy@philkeandesigns.com

Casa Siena
Web: www.casasienadesign.com
Email: info@casasienadesign.com

Stephen Ryan Design & Decoration
Web: www.stephenryandesign.com
Email: info@stephenryandesign.com

Casa Forma Limited
Web: www.casaforma.co.uk
Email: info@casaforma.co.uk